Automating Governance in China?

AUTOMATING GOVERNANCE IN CHINA?

Data-Driven Systems in the Scoring Society

Edited by

Haiqing Yu and Rogier Creemers

Leiden University Press

Cover design: Andre Klijsen
Cover illustration: © BirgitKorber
Lay-out: Crius Group, Hulshout
Printer: Scanlaser bv, Zaandam

ISBN 9789087284657
e-ISBN 9789400605022 (e-PDF)
e-ISBN 9789400605428 (e-PUB)
https://doi.org/10.24415/9789087284657
NUR 740

Contact EU General Product Safety Regulation (GPSR): productsafety@lup.nl

Table of Contents

List of illustrations and tables

Acknowledgements

This book project was funded by the Australian Research Council (ARC) Future Fellowship (FT200100100). It was also supported by the ARC Centre of Excellence for Automated Decision-Making and Society (ADM+S), the CatCh Network (Aalborg University, Denmark), and the Netherlands Organisation for Scientific Research (NWO) Vidi project "The Smart State: Big Data, Artificial Intelligence and the Law in China" (016.Vidi.185.200). With their support, we were able to workshop selected papers at the "International Conferences on Automated Decision-Making and Chinese Societies" (1–3 February 2023, RMIT University, Melbourne) for this book.

Our heartfelt thanks go to all the contributors to this volume. We thank them for their collegiality, patience, and cooperation. It's been a pleasant journey working on this book project.

As always, we thank our families for their unwavering support and love.

Haiqing Yu and Rogier Creemers

CHAPTER 1

Automating Governance and Data-Driven Scoring in China: A Critical Introduction

Haiqing Yu and Rogier Creemers

Abstract

This chapter offers a critical introduction to key concepts, central questions, and the scope of this volume. It reviews how China has encouraged and employed automation technologies and toolkits to enhance social governance. Central to the discussion about automation and governance is the role of data in tracking, sorting, dividing, categorizing, and scoring practices. The chapter discusses examples and discontent in China's data-driven governance and its datafication and dataveillance practices in its digital innovations. China offers a reference point for rethinking the politics of automation and data-driven systems being used in governance in the 21st century. The global dimension of China's informationization, digitization, and automation in transforming and reshaping our political systems, socioeconomic fabrics, and cultural ethos warrants nuanced, contextual, and comparative analyses from multidisciplinary researchers. By not merely contrasting China to Western expectations but considering it as a starting point in approaching questions of technology and society, this book intends to contribute to a more inclusive and culturally diverse approach to scholarship on automated decision-making and society.

Keywords: Automating governance; Social governance; Regulatory state; Data; Scoring; Global digital China

Introduction

Perhaps more than any other government in the world, the Chinese government is embracing the potential of artificial intelligence (AI), blockchain, and other data-driven technologies in governing its society and economy. Over the past decade, it has released a series of major policies and plans on using big data (State Council 2015), artificial intelligence (State Council 2017), and blockchain (Yang, Yang and Liang 2022) for constructing a "digital China." (Central Committee and State Council 2023). In the 14th Five-Year Planning cycle, a dedicated document provides a blueprint for the overall digitization of government services (NDRC 2021). This development has seen a terminology change in China's official documents written since the 18th congress of the Chinese Communist Party (CCP) in November 2012.

When referring to social control and management the language has shifted from "social management" (*shehui guanli* 社会管理) to "social governance" (*shehui zhili* 社会治理). More recently, "digital" has been added as a modifier to social governance, as in "digitized governance" (*shuzihua zhili* 数字化治理) and "digital transformation of social governance" (*shehui zhili shuzihua zhuanxing* 社会治理数字化转型) (Li and Wu 2022; Xiang 2021). It is required that the digitized social governance system is led by the CCP, coordinated by the government, participated in by social actors and the public, protected by law, and sustained by science and technology. In other words, social governance is centrally controlled and coordinated—by what Xi has termed "top-down design" (*dingceng sheji* 顶层设计)—and enabled by an evolving digital, smart, new type of infrastructure. Science and technology, known a century ago as "Mr Science" (sai xiansheng 赛先生) during the May Fourth Movement in 1919, is again called upon to usher in a new era of digitization (*shuzihua* 数字化 and intelligentization (*zhinenghua* 智能化).

AI is a key component of efforts to build the new type of infrastructure, networked and data-centred, to enhance China's productivity and prosperity (China Information Centre 2021). The judiciary is experimenting with AI and big data to enhance its efficiency and reduce the workload of its overburdened and under-resourced staff (Papagianneas 2022). Increasingly powerful algorithmic sorting and facial recognition tools are incorporated into growing surveillance networks, as China's ever more innovative and sophisticated tech companies continuously improve their products in the hope of attracting lucrative government procurement contracts (Batke and Ohlberg 2020). Tech giants such as Huawei, DJI, and Hikvision supply a range of hardware and software to equip China's massive surveillance networks, while Baidu, Alibaba, and Tencent (BAT) are contractors and operators for e-government services (from city services to the health code), as well as for consumer-oriented third-party businesses of public and private suppliers (including of services such as payment, travel and transportation, shopping, and entertainment). The private sector has enthusiastically adopted data-enabled tools for a range of purposes, from targeted advertising and content recommendations to work assignment algorithms for delivery drivers. In short, Chinese individuals are increasingly confronted by a context in which both governmental and corporate actors are engaging in automated decision-making (ADM).

The nature and applications of ADM tools and technologies are highly diverse. They differ in technological terms. Some applications require very sophisticated forms of AI, machine learning technologies, and blockchains, as well as very large sets of training data; others—such as some functions of the social credit system—are far less technological, employing traditional bureaucratic data collection and entry methods with the help of Weixin (China's all-in-one super app). They differ in terms of functionality. Some systems, such as the health code system, are aimed at scoring and sorting individuals based on sets of behavioral parameters, relational, temporal, and spatial, with consequences flowing from that evaluation as in China's COVID-19 pandemic control (2020–2022). Others, such as the application of facial recognition technologies in policing and surveillance, are intended to facilitate citizen identification and community safety. Apart from technology and functionality, they also differ in their purposes. Government agencies tell us that they use ADM tools and systems to improve efficiency and thus achieve more coordinated social and economic governance. Companies are reported to have used these technologies to streamline workflows and maximize revenue.

There are different types and levels of ADM in practice. Ulrik Roehl (2022), for example, offers a three-level, six-level typology in administrative ADM, from "no automation" and "semi-automation" (acquisition and presentation of data, suggested procedural steps, supported decisions) to "full automation" (automated decisions and autonomous decisions). He cautions against technologically deterministic understandings of ADM, calling for contextualizing of technology in user experiences and cultural milieus. We heed his advice about contextualizing and localizing ADM systems including their failures (see Chapter 2). That means the term "automation" will be used in a broad sense in this book. On the one hand, it can include applications in social governance systems that do not necessarily contain sophisticated digital decision-making tools. On the other hand, it may also include highly targeted price discrimination algorithms used by e-commerce platforms (see Chapter 9).

Moreover, the deployment of ADM tools and technologies is taking place within rapidly changing socioeconomic and geopolitical contexts. Concerns around surveillance, for instance, have been a major impetus for justifying the imposition of US export sanctions against China, and for foreign governments to review the domestic use of China-sourced surveillance technologies. Both Chinese citizens and its government are increasingly aware of data-related privacy concerns reflected in the promulgation of the Personal Information Protection Law. This law seeks to navigate the tension between ensuring autonomy over the collection and use of data pertaining to individuals while disciplining data-handling activities within government departments on the one hand and enabling police and security services' operations on the other. Lastly, it is also feeding into learning processes in government itself.

In sum, China offers a reference point for rethinking the politics of automation and data-driven systems being used in governance in the 21st century. The country has embraced AI, blockchain, and big data technologies as key drivers of economic growth, technological innovation, and digital governance. It offers excellent case studies through which to examine the technological, sociocultural, and geopolitical dimensions of digital strategies in governance innovation. It also offers a comparative framework for examining the role of automation technologies in reshaping state-society relations. In this introduction, we highlight two key concerns that run throughout this volume, concerns regarding the automation of governance in China and, central to it, the role of data-driven scoring practices.

Automating Governance in China

China's journey toward digitizing and automating governance started as early as the 1950s with its attempts at simplifying and automating Chinese characters in typing, printing, and computer inputting (Tsu 2022). As Tsu has pointed out, through the technological revolution of the Chinese script China has not only caught up with developed countries in technological and economic development but has also restored its confidence. The only dream it is now chasing is the digital future that it envisions as the "Chinese dream" and "community of common destiny" (buzzwords of Chinese President Xi Jinping), often portrayed in the West in dystopian terms (by Hayes 2020, for example). Although China has a long history in its experimentation with digital and automation technologies in social governance, in the Xi era it has taken a great-leap-forward approach to digital governance.

For decades, the CCP has cultivated a technocratic approach to governance. From the 1980s onwards, it has married the Marxist principle—that the historical evolution of human societies progresses along a scientifically predictable path—with an eclectic set of ideas derived from cybernetics, systems theory, and social management theory (Knight 2025). Governing, defined as implementing a science-based programme for human progress, thus became intimately intertwined with social science research on the assumption that, if more data was collected and better analyzed, scientifically optimal policies could be derived. Similar to the Webb Space Telescope enabling greater knowledge about the universe, the Chinese government believes that more data and greater computational power provide the instrumentation for deepening its science-based governance capabilities and realizing more of the promise of its theoretical framework for achieving progress. In the government's view, digitization carries many benefits. Automating governance may reduce, for instance, much of the discretion (and thus corruption) of local officials, either by removing decision-making power from them or by exposing them to greater scrutiny.[1]

[1] This is also a goal of the "rule of law" reforms implemented over the past decades. In official texts, "rule of law" (fazhi) is often juxtaposed with "rule of man" (renzhi), suggesting that the leadership intends it to reduce human inconsistency rather than to impose meaningful legal constraints on state action as the Western reading of this term implies.

Yet the transformation of these broad ideas and ambitious goals into actual, concrete policy, regulation, and governance practice raises many questions about both ADM systems and their inclusion in the state's operations, and about the broader reforms and transformations in the relationship between the state and citizens that they enable. Moreover, such transformations are part of dynamic processes in which lessons learned from earlier measures, new technological advances, and developments in private industry continually create new feedback loops and impetus for change. It would therefore be beside the point to try to identify a specific "Chinese" approach to ADM at this stage. Instead, many insights can be gained by taking a more granular approach, focusing on specific ADM practices, applications, or discourses, and connecting this micro-perspective to macro-level questions both about China and about the evolution of ADM technologies worldwide.

In this sense, this book follows the dominant tendency in the literature to analyze the Chinese deployment of technologies through the perspective of authoritarian governance, focusing on discourses and applications in AI, blockchain, and other ADM technologies and systems in surveilling, profiling, categorizing, and servicing the world's second largest population. China is, of course, not a liberal democracy, and no other government in the world is as ambitious in automating its society as is China's—it is unique in that sense. However, many topics and approaches are missed through the spotlight beam on authoritarian governance in the current literature. That focus tends to foreground an antagonistic state-versus-society perspective, for instance, paying far less attention to the complexities within the CCP or the Party-state itself, or to emergent techno-political landscape and multipolarity in modes of articulations via blockchain technologies (see Chapter 3). Even if it is the goal of the CCP to retain its hold on power (and there are few incumbent political organizations or individuals who do not), that tells us little about the evolving ideas and practices underpinning the realization of that goal; these are the subject of relatively diverse debates, particularly on how to configure and manage new data-driven ADM systems at national and local levels.

China's approach to automating governance is state led, driven by a long-term strategic vision for national development and a desire to maintain information control and social stability. Most of the research on digitization in China has focused on questions of political risk-related social control, including surveillance and stability maintenance. This reflects growing anxiety about the "perfect dictatorship" (Ringen 2016) of the Chinese Party-state and its increasing digital capabilities in social and opinion management (Creemers 2017). Indeed, digitization in the judicial reforms, for example, reinforce the Party-state's authority over the judiciary while seemingly contributing to professionalism in the courts (see Chapter 4).

Lastly, regime type-based studies tend to overestimate the degree of centralization and coordination while underestimating the messiness that characterizes realities on the ground. It bears remembering that the CCP counts more members than Germany has citizens, and that it governs over thousands of large cities and small towns. Activity at the local level is, in many ways, at least as impactful as that driven by the central government; experiments and pilot projects take place at that level and are later expanded across the entire nation—it is where central directives must be transformed in the light of local realities.

In a People's Daily article in response to the CCP's 20th congress (held in October 2022), Chen Yixin (Minister of State Security) laid out three core objectives in "improving the social governance system": *political security* (referring to terrorism; foreign interference, and sabotage), *social stability* (*weiwen* 维稳, such as collective incidents and petitions), and *public safety* (including crimes and workplace safety). He emphasized local and grassroots social governance—so-called "bottom-up logic" in response to "top-down design"—to promote the modernization of social governance (Chen 2023). Such a coordinated approach to social governance must be supported by "intelligence", that is, data-centered, networked information databases, such as the Integrated Joint Operations Platform, a massive database combining personal data automatically harvested online from public and private platforms together with information that is entered manually by on-the-ground "grid" members.[2] Also known as China's "Big Brother App" (Wang 2019), the platform is one of many policing and social governance platforms that require "boots on the ground" (grid management) or other work of humans in the loop to support the automation of governance.

Data-Driven Scoring in China

Central to the discussion about automation and governance is the role of data in tracking, sorting, dividing, categorizing, and scoring practices. China has been known as a scoring society with a long history of categorizing and ranking its population through policies or decrees and conventions (Ghosh 2020; Wallace 2023). As von Galahn (2012, 39) writes, "[c]ivil registration for the purpose of social control and the mobilization of labor and other resources was a cornerstone of the Chinese imperial state." Information-gathering about, and registration of, individuals, households, clans, and their assets and property was an important tool of governance in China's long and continuous history of civil registration in

[2] Piloted since 2004 to improve grassroots management, China has revived and optimized the grid governance practice by introducing intelligent, automatic control and communication mechanisms to enable information-sharing and data integration. It has proved pivotal in the maintenance of China's stability and its COVID-19 lockdown management. See Mittelstaedt (2022), for example.

successive dynasties for the purposes of taxation, social reproduction, labor service, military conscription, and land allocation. It enabled rulers to manage and discipline their society and to impose military-style organization and mobilization in times of crisis.

Central to the civil registration is anchoring subjects to their registered residence. The Chinese household registration system, known as *hukou*户口, is not a completely new paradigm of Chinese socialist modernity or an exclusive invention of the CCP. For centuries household management tactics, such as the *baojia* system, have been used "to address the informational needs that result from a relatively high labour-land ratio" in a relatively densely settled agrarian economy (Szreter and Breckenridge 2012, 24). Hukou has been used to divide the Chinese population into urban and rural groups. It is used like a caste system under Mao to strictly control not only population mobility but also resource allocation. In the post-Mao era, the hukou system has been maintained but internal migration restrictions are loosened. This has created a "floating population", with millions of young villagers migrating into the cities to work in construction, export-oriented manufacturing, household service and other jobs that urbanites would look down upon. Meanwhile, cities like Shenzhen have used the *hukou* mechanism to attract "talents"—the so-called high-quality (*gao suzhi*) Chinese citizens—by designing a points system on the basis of a numerical assessment to "scientifically" allocate citizen rights and public goods (see further discussion in Chapter 7).

Although its purpose and practice have undergone dramatic changes over two millennia, *hukou* in the PRC era has continued its long-lasting trajectory in China's data-driven scoring governmentality through scoring, ranking, and sorting. In the post-reform era, there have been various attempts to implement innovations in this tradition of population control by making it more scientific and accountable. This follows the global trend to adopt the data logics, technologies, and automation processes of the private sector into the practices of citizen scoring and techno-social shaping of citizenship (Dencik et al. 2019).

Citizen scoring refers to "the use of data analytics in government *for the purpose of categorisation, assessment and prediction at both individual and population level*" (Dencik et al. 2019, 3; original emphasis). It is emblematic of the logics of data-driven scoring and rating in contemporary societies, from financial and commercial industries to governmental and public services. China's social credit system is often used as an example of data-driven governance and state surveillance infrastructure (Liang et al. 2018; Backer 2019). The system is best known for its credit scoring and ranking, with low-scoring individuals and corporates being placed on blacklists (the untrustworthy categories) and those with higher scores on the red lists (trustworthy categories), thus determining different kinds of preferential treatment or punishment and cultivating an ideal, loyal citizenry (Tsai et al. 2021; Hou and Fu 2022). Implemented by local governments, earlier attempts to embed the social credit system (SCS) in the social and cultural fabric of everyday life failed, often resulting from a backlash of popular discontent and protest. The SCS is predominantly limited to the financial sector and used as a compliance tool (see Chapter 6).

As the social credit system is normalized as an evolving method of social control rather than as the West's imagining of dystopian authoritarianism (e.g. Creemers 2018; Yu 2023), it is easy to see its relevance to the discussion about what David Lyon (2002) calls "social sorting." Social sorting "highlights the classifying drive of contemporary surveillance" and "defuses some of the more supposedly sinister aspects of surveillance processes" that "not only rationalize[s] but also automate[s]" the process of social and personal categorization (Lyon 2002, 13). Whether it is in marketing or policing, sociotechnical surveillance systems have turned people into data subjects and data doubles who are searchable via databases, on the move, and modifiable. Using the language of "efficiency, productivity, convenience, and comfort" (ibid, 18), covert practices of social sorting appear benign and innocent. In the Chinese context, social sorting takes a "pan-moralism" tendency (Bakken, 2000); in our present discussion this pan-moralism can be described as a tendency to ground arguments in morality discourses, whether in regard to the reasons for corruption and fraud or the reasons for poverty and disorder (Yu 2008).

Social sorting shapes moral conduct, disciplines subjects, and creates social stratifications based on a wide range of vectors in human variables. It exercises power over the social body when empowered by data assemblages—including digital transactional records, video images captured via facial recognition cameras, biometrics (fingerprints, iris scans, or DNA samples), geolocation tracking technologies on mobile phones, and computerized administrative files—in the name of building a trustworthy society and good citizenship (Liu, Lin, and Chen 2019; Zhang 2020). Using legal, administrative, and technical means, huge amounts of data can be collected and processed quickly to inform decision-making in social control and governance, such as policing the LGBTQIA+ activists (see Chapter 8).

During the COVID-19 pandemic, the Chinese government made epistemological claims to scientific truth and methods in pandemic control based on its ability to mobilize all public and private resources in data collection and processing. Facilitated by popular social media platforms like Weixin and AliPay and empowered by big data, facial recognition technologies, and geolocation tracking technologies, the health code system was regarded as the magic weapon in China's "success" in combating the pandemic (Yu 2022). While the health code system is now no longer used, it has never been officially abolished—it remains in people's mobile phone apps. What has been monitored and traced is not merely where people have been and what they have done, but also with whom they interact or are connected, and how long they engaged with particular people, products, or services. Such information feeds into the datafication machinery of surveillance capitalism and digital authoritarianism.

Fourcade and Gordon (2020) argue that modern bureaucracies derive their power from information via their vast sociotechnical machinery; public and private actors and interests comingle to "mint" data and create a new form of governmentality called the dataist state (or machine learning state and artificial intelligence state). Such a dataist state exercises data-driven social, economic, and political control over individual and social bodies, often in collaboration with corporate interests (as in value extraction such as via algorithmic price discrimination), and seeks monopoly over population classification and resource distribution. The power of the dataist state is like the rhizome, with ubiquitous roots, shoots, and modalities; it is omnipresent in the social body yet lacks public accountability for its actions or for its underlying structural inequalities, stigmatization, and biases.

The examples of the social credit system and the health code system reflect a dominant ideology in China regarding the teleology of technology and the technological fix. Both systems use big data-driven social sorting and scoring technologies to modernize bureaucratic capability and to govern and organize social and economic activities. Such a trend reflects a deep-rooted desire among Chinese elites for scientific management practices rooted in "technoscientific reasoning" (Sigley 2009). Technoscientific reasoning derives from "knowledges concerned with shaping human conduct" based on modern claims to "scientific truth" and reflects "the desire within governmental and administrative projects to create certain human subjects" (Sigley 2009, 538). It derives and exercises its power from and upon the individual and social body "in order to shape human conduct and thereby forge new relations between sovereign and subject, between nation and citizen" (ibid, 542).

The devil is always in the detail, but this is not a place to discuss in detail China's role in surveillance capitalism or its surveillance industry. The explanation of China's obsession with data, scoring, and surveillance lies as much with Xi Jinping and George Orwell as it does in Shoshana Zuboff (2019) and Kaifu Lee (2018). It is worth pointing out, by way of Cohen (2019), that surveillance capitalism ensures that internet platform firms have control over individual data and that platforms successfully promote a culture of access-for-data and acceptance of data-driven algorithmic service methods among users by accepting cookies or terms of service for using online services. This is despite global anxieties about surveillance capitalism, which underpin current debates on the turn to the state and its role in content regulation (Flew 2019). China has led the trend for using state power to regulate digital platforms. Since 2020, Chinese regulators have imposed a series of measures intended to better protect the personal information of online platform users and limit the degree to which they can exploit data for economic gain, thus effectively erecting significant obstacles and limitations to the data-driven "surveillance capitalist" business models of China's digital giants (Creemers 2023).

The regulatory state worldwide increasingly adopts compliance monitoring and reporting mechanisms that are intensively managerial in orientation, outsourced to specialized professionals, and seemingly automated. The situation creates regulatory fragmentation, oversight, and even performative regulation for the sake of appearances. The ubiquity of social control and surveillance tools and systems in China—from what one can see in regular use (e.g. surveillance cameras equipped with the facial recognition technology, see Chapter 5) to what is less visible but no less controversial (e.g. queer social sorting, see Chapter 8)—does not equal total control, as many of the technologies and systems do not always work as intended, and automation is not implemented at the local and grassroots levels (see Yu 2019, for example).

Most research has focused on critiquing elite or bureaucratic enthusiasm for data-driven governance. Some have pointed out the paradox of technocracy and the aesthetics of "open" data in various smart city initiatives. It is known that after more than two decades of e-government initiatives, smart government is limited to bureaucratic efficiency and is unlikely to translate into the politics of open government (Yu and Robinson 2012). It is also known that bureaucratic knowledge and capability at grassroots levels often fall short of expectations. Local cadres and grid members are too buried in data entry and reporting to carry their jobs into the streets, and they would fake records in order to meet their targets, or resort to sending manipulated information into the big data system to con officials higher in rank (Bakken 2022). They are also prey to power struggles within the power-money-intellect iron triangle (Zhao 2008). This is seen in the Henan health code scandal in which the metadata of certain individuals were manipulated to stop them gathering in public to protest and petition, all in the name of stability maintenance and pandemic control (Yu and Zeuthen 2023). Rather than being subject to criminalization, local cadres are demoted or given demerits according to CCP's internal disciplinary policy. They are made scapegoats of a discriminatory system designed and promoted by the state and its willing partners in the private and public sectors.

Anyone who has visited China recently will have noted the omnipotent ID card in the lives of ordinary Chinese citizens, for banking, ticketing, hotel registration, online account registration, and many other public and private services. The second-generation ID card, referred to as a "smart card" (read by any radio-frequency-identification device), has an embedded digital microchip that contains the cardholder's information, including name, sex, birth date, address and *hukou*. Known as the e-ID card, it has gone digital, and has been piloted in more than 15 major cities since 2018; it enables people to use facial recognition technology on Weixin to verify and authenticate their identity. The government has called for acceleration of the application of new technologies, including big data, cloud computing, and AI, in the e-ID card system (Zhong 2022).

The e-ID card represents the nexus of data collection both online and offline. It is the point of entry of scalable databases controlled by the Chinese government for policing, surveilling, and managing population mobility, and for allocating resources and services. As Chapter 7 will discuss, *hukou*, on the other hand, became a marketable commodity when provincial and municipal governments introduced the points or scoring systems to allow "outsiders" (non-locals) with wealth to obtain the right to live in metropolitan cities like Shanghai or Shenzhen. Its role as a status barrier or discriminatory system leads to social inequality and polarization on a massive scale; it dichotomizes the population into urban and rural, underclass (the so-called low-*suzhi*, "low-end population") and upper class (the so-called "high-net-worth individuals"), and into information and rights haves and have-nots (especially in terms of access to social services—see Qiu 2009; Wang 2019); it creates a regime of hierarchical difference, differentiating citizenship, rights, mobility, and lived experiences in relation to geographical, economic, gender, employment/professional qualifications, and CCP membership matters, as well as in wealth/economic class through points-based *hukou* status. *Hukou* has continued to perform its role in dividing, categorizing, and managing population mobility and resource allocation in China's form of "apartheid" (see, for instance, Mallee 1995; Wang 2005; Borge 2022). Together with *suzhi*, *hukou* becomes a technology of governance (Zhang 2023). It is now embedded in the national ID system, which is in turn embedded in the social credit and other social governance systems.

The datafication, categorization, and scoring of population goes beyond the *hukou* or the e-ID card system. It is part of the data-driven social governance craze that continues the PRC tradition of identifying and separating targeted groups of people for control and surveillance purposes. From class enemies in the Mao era to criminals, political dissidents, and the floating population in the Deng era, and to low-end populations, Muslim separatists, anti-China (*fahua* 反华) or insulting-China (*ruhua* 辱华) "enemy of the Chinese people" in the Xi era, Chinese authorities have been very much preoccupied with governing "outsiders" in their carefully built "harmonious" systems.[3]

Such systems are used not only to score and rank people in the social hierarchy as determined by wealth (as in the points-based *hukou* system), morality (as in queer social sorting), or political loyalty (as in the smart courts), they are also used to evaluate the performance of government officials, including the under-paid and overworked grid members, neighborhood committee members, urban

[3] We use the term "harmonious" to refer to the legacy of the Hu-Wen era, described as a "scientific development concept". See for example, Delury (2008) and Geis and Holt (2009).

management teams, and their subcontractors. Cadre evaluation and promotion, as Wallace (2023) points out, is part of China's limited quantified system of governance. The top-down bureaucratic management system asks local officials to resolve their own local problems and suffocates any kind of grassroots engagement (Mittelstaedt 2022). Cadres are rated, rewarded, or punished for their economic performance as well as their *weiwen* performance.

Everybody, except the ruling elites (including the Politburo members of the CCP), is scored, including local officials who implement centrally designed schemes and community workers who collect individual data and monitor residents within their jurisdictions. People are color coded—green, yellow, or red in the health code system— and put on color-named lists, the red list or blacklist in the social credit system, while also being ranked into grades—A, B, and C in cadre evaluation or for predictive policing (Bakken 2022). Quantification conveys a sense of objective truth, scientific authority, and managerial efficiency while hiding the structural or systemic issues driving exploitation and inequality and stifling any public discussion of the data-driven scoring systems in China's digitized and "automated" social governance.

Theoretical Contribution

This book not only intends to collect empirical insights into Chinese ADM processes; it also speaks to broader emerging concerns in the field of science and technology studies, where China presents an insightful case study supporting global theory-building (Franceschini and Loubere 2022). With algorithmization, datafication, and automation, social governance has seen increasing collaboration between governments and tech companies (including social media platforms) in expanding ADM in the social sector. Digital governance has become more powerful, intrusive and pervasive, while at the same time it is supposed to be more responsive and inclusive (Katzenbach and Ulbricht 2019). Questions about how data is generated, collected, used, and governed by public and private actors have generated increasing interest and imagination among academia, media, and government agencies (Dencik et al. 2019). This is especially pertinent when discussing the Chinese government's use of datafication and dataveillance in its digital governance innovations.

To build smart governance systems, the Chinese government has been working closely with state-owned and private telecommunication and digital companies to use new and emerging technologies to improve administrative efficiency, public service delivery, digital economy infrastructure, and consumer connectivity and experience. China's digital strategy is multi-dimensional. It is *technological*—ADM technologies and systems, comprising an array of intelligent and emerging technologies from artificial intelligence, machine learning, to blockchain, are used to innovate social governance, service provision, transport/mobility, and knowledge production across many sectors. It is *sociocultural*—the everyday experiences and processes of ADM technologies and systems have practical and philosophical implications for individuals and communities in the era of digital transformations. It is also *political*—the development and adaptation of ADM systems, before, during and after the COVID-19 pandemic, highlight the role of information technologies and big-data governance in China's digital transformation and sociotechnological imaginary of the future.

China's approach to automating governance is state led, driven by long-term planning, a strategic vision for national development, and a desire to maintain information control and social stability. Most of the research on digitization in China has focused on questions of political risk-related social control, including surveillance and stability maintenance. This reflects growing anxiety about the "perfect dictatorship" (Ringen 2016) of the Chinese Party-state and its increasing digital capabilities in social and opinion management (Creemers 2017). However, this comes at an analytical cost. First, this focus on "authoritarian teleology", where every governance action is explained in the light of maintaining power, obscures many other motivations and impedes a comprehensive understanding of the inherent functioning of the system (Kuo 2024). Furthermore, it may limit our ability to include China into broader theory building. From a comparative perspective, consigning China to the "authoritarian" bin implies it bears little or nothing in common with the democratic state, thus diminishing our ability to understand evolutions in China as part of a broader phenomenon of integrating ADM technologies into governance and commercial processes—with features shared across countries regardless of their regime type. Surveillance is, after all, endemic to the business models of any large online firm (Zuboff 2019). As pointed out earlier, regime type-based studies tend to overestimate the degree of centralization and coordination and underestimate the messiness and differences that characterize realities on the ground.

The ramification of digital strategies taken by nation-states is not bound by any geographic boundaries. New forms of ADM systems are experimented in China—often pioneered by its tech giants and with transnational footprints in financing, research and development, and marketing. China has seized the opportunity to leapfrog from being a follower to a competitor and leader in the design, control, and applications of ADM technologies and systems across a wide array of areas and sectors. It is also competing with Western (American) powers to control a huge amount of data internationally, and by extension resources, ideas, intelligence, and power. China's digital sphere of influence is not limited to the PRC or the Chinese-speaking communities around the world with China-centric digital technologies, platforms and infrastructure (Keane et al. 2020; Sun and Yu 2022). Chinese surveillance technologies have been exported around the world (Mozur et al. 2019).

In the spirit of global digital China, this book calls for a multi-dimensional, comparative perspective to examine the changing innovation and technological landscape and its discontents within the country. As Peter Yu in Chapter 10 points out, just as we pay greater attention to automated governance and the use of algorithms and ADM systems in the country, we can also benefit from a deeper appreciation and understanding of the strengths and weaknesses of different approaches to technology innovation and regulation taken by the European Union, the US, and other parts of the world, by placing China in the world, rather than singling it out as an exception or the other. Questions like how to manage multivariate divides in the face of algorithmic dominance and the human–machine interface are so novel and technologically challenging that we all struggle to answer them: all countries, stakeholders, and individuals are scrambling to make sense of new and emerging buzzwords; they develop guidelines to help make difficult choices and safeguard humanity. Such comparative analyses will allow us to better appreciate the case studies in this book and better prepare for the unintended consequences and spillover effects sparked by technological developments and policy choices.

It is our hope that by outlining many possible scenarios and directions in China's engagement with new digital technologies and technological systems to optimize and automate social governance, this book will encourage academic and policy researchers to devote greater attention to and more nuanced perspectives on the experimental nature and processes in societal applications of technological innovations. Taking global digital China as a method not only entails an appreciation of the country's integration with the global digital capitalism and therefore its seemingly parallel development vis-a-vis the West in response to changes precipitated by advances in big data analytics, machine learning, AI, and automation technologies, but it also implies an ideological shift in focus, from a bipolar to multipolar world where new universalities grounded in specific spatial-temporal logics are possible.

The Book: Objectives and Overview

The book speaks to increasing global interest in China's rise as a digital superpower and contributes to understanding China's instrumental and pragmatic approach to the use of disruptive technologies to reposition itself as the Central Kingdom in the global "community of shared future" (Nathan and Zhang 2022). It aims at a holistic understanding of the relationship between digital technology and technology of governance in China while seeking to illuminate how disruptive technologies are pivoted to the Chinese state as it pursues development and seeks new sources of legitimacy while transitioning into middle-income status. This brings new economic and social challenges. The book addresses two key questions: 1. What are the forms and features of automated decision-making systems operating in China? and 2. How does China's experimentation with automation technologies (via datafication and algorithmic governance, for example) inform us regarding the relationships of power, knowledge, and ordering of social life?

This edited volume includes ten chapters, written by established scholars as well as early-career researchers doing cutting-edge critical research into the social and legal implications of science and technology in the context of China studies. The chapters address China's digital strategy in social governance domestically, and its geopolitical implications internationally. They review the policy and regulatory frameworks informing the vast range of digital applications under development. They also offer case studies of specific projects and applications that are automated or semi-automated in organizing and governing social, economic, and cultural lives in China. Case studies illustrate new modes of digital governance and authoritarianism conceived by the Chinese government as it interacts and collaborates with technology companies, the Chinese people, and other stakeholders. They offer new insights into the deployment of ADM in Chinese governance, and the way in which it is applied and implemented in real-life scenarios.

The chapters highlight a diversity of topics that present a more complex picture than the state-versus-society paradigm or surveillance framework might suggest when digital technologies and China are discussed. They include commentary on how the use of ADM may be intended to forestall the emergence of professional classes with autonomous identities and self-perceptions, or to enhance the functioning of state bodies, but also on how the failures and shortcomings of ADM are perceived within the system, and on how China seeks to overcome the digital divide by expanding connectivity and access. Written by a multidisciplinary group of scholars, the chapters integrate multiple dimensions—the technological, social, cultural, political, and in some cases global. They consider the interplay between the affordances of technologies including AI, machine learning, and blockchain, the experiences and processes of ADM systems and the actors affected by them, and the process of learning and adaptation by state actors as part of governance reform.

The ten chapters are organized into two themes: discourses and applications. They examine dominant discourses on Chinese ADM technologies and systems, from ADM failures and algorithmic divide to blockchains in governance and digitization of the judicial system. They also provide case studies in ADM applications, from the facial recognition technology, the social credit system, and the points-based *hukou* reform to queer social sorting and algorithmic price regulation.

Following this introduction chapter, the book starts with Beijing-based legal scholar Xin Dai's chapter on "Locating and localizing automated decision-making failures in China". It uses a heuristic typology to examine the diversity of real-world ADM failures, ranging from health code debacles to credit reporting for small and medium-sized enterprises, from algorithmic recommendation systems to gig work platforms' assignment systems. With reference to such examples drawn from China's experience with deploying ADMs, Dai cautions against the simplistic application of Western perspectives, attitudes, and discourses on ADM to the descriptive and prescriptive thinking about ADM failures in non-Western cultural and social contexts. Such localized perspectives on ADM systems and their failures will enable societies to better harness the decision-making advantages of ADMs while living with their inevitable shortcomings.

In Chapter 3, "A democratic ethos? Explorations of blockchains and governance in China," Australia-based political scientist and blockchain researcher and practitioner Warwick Powell discusses blockchain technologies in China's economic and social governance. Powell draws on insights from Ranciere and Durkheim to discuss *data ecologies* and to contextualize his focus on blockchain technologies in China's new mentality of governance. The chapter suggests that blockchains with Chinese characteristics are the emerging foundations of a potentially new ethos of socialist governance. Application of this ethos hinges on the expansion of the public sphere of data via the creation of data ecologies anchored by blockchains, and the capacity of symmetric and synchronous data to sustain common truths essential to the dialogue that makes functional democratic governance possible. Chinese ambitions for governance reform, namely those of co-production, co-governance and co-sharing (*gongjian*, *gongzhi* and *gongxiang*), are amplified through the capability of blockchains to enlist a plurality of agents to a common informational integrity purpose, a purpose that is necessary to sustain the deliberations and reflections that take place in consultation between the governed and the governing.

Chapter 4, which is authored by Germany-based researcher and legal scholar Fan Yang, is entitled "Techno-utopia or techno-trap? Unveiling the enigma of smart courts in China's judicial reform." The chapter explores the impact of China's smart courts on diverse judicial reform measures. Through studies on four provincial data-driven systems in Chinese courts, it illustrates the paradox of smart courts as a limiting factor and even a hindrance to judges and local courts in reforming China's judicial system to achieve efficiency, consistency, procedural compliance, and the means to address misconduct. Ultimately, smart courts reinforce the Party's authority and control over the judiciary and its actors through the automation of multi-layered party oversight mechanisms.

Chapter 5, written by five social science scholars based in Australia and led by cultural and media studies scholar Xin Gu, examines the public response to and the discourses on facial recognition technology (FRT) in China. Entitled "Balancing control and engagement: China's sociotechnical imaginary in facial recognition technology," the chapter argues that the deployment of FRT extends China's sociotechnical imaginaries by offering individuals a mechanism to articulate their concerns and resistance at the grassroots level. Despite the government's oversight, the technology inadvertently opens channels for public dissent and negotiation. Through a case study of a micro sociocultural context of a middle-class housing compound in Shanghai, the chapter reveals how China's approach to automated surveillance technologies like FRT must continually reconfigure the delicate balance between social control and civic participation. It also considers how the pervasive use of FRT for mass surveillance may destabilize this fragile equilibrium, posing potential risks to the state's ability to maintain both governance and public trust.

Chapter 6, contributed by US-based political scientist Haemin Jee, focuses on local implementation of the social credit system. In her chapter, "The social credit system as a law-enforcing tool: pillars of local implementations," Jee investigates how the social credit system is implemented by local governments and uncovers new insights regarding its uses. First, local social credit systems mainly target and punish firms rather than individual citizens. Second, these targeted firms are punished under the social credit system for violating existing laws and regulations, behaviors that were already deemed unlawful and subject to punishment by existing legal and regulatory frameworks. These core findings reveal a surprising function of the social credit system—it can be used to reinforce existing institutions and garner more compliance with *existing* laws and regulations from firms, rather than constricting and monitoring individual-level behavior.

Chapter 7, from France-based China studies scholar, Anne-Christine Trémon, continues the discussion of local implementation of data-driven governance, examining China's embrace of "smart governance" and e-government via the points system. Entitled "Scientific fairness: justifications and critiques of points systems in Shenzhen," the chapter critiques the justifications that designate points systems as a scientific governance tool that ensures fairness in allocating rights to urban citizenship or to locally provided public goods on the basis of a numerical assessment and ranking of applicants. This discursive legitimation invokes not only their scientific but also their "reasonable" character. This shift from "scientific" to "reasonable" makes "scientific fairness" understandable among the public, opening up a space for potential critique; and yet, the justifying adjectives of "scientific" and "reasonable" act both as tools of critique and as tools for shutting down critique.

Chapter 8, "Queer social sorting: control and criminalization in China's LGBTQIA+ activism," is contributed by Australia-based security and surveillance studies scholar Ausma Bernot. Following the earlier discussion of social sorting, Bernot discusses how LGBTQ+ activism in China has been progressively hampered by sophisticated queer social sorting methods that include both automated decision-making and the use of human labor in enforcing queer "othering." She points out that the suppression of LGBTQIA+ activism exists under the thin surface of political decorum and formulaic support for such communities in formal announcements confirming China's stance against discrimination as well as the gradual thawing of national laws that restrict sexual minority communities. The state-sponsored control of queer activism is now increasingly linked to two elements: the call to return to traditional gender roles, and *social sorting* of queer activism. Queer social sorting is achieved through the interconnected tools of legal and regulatory frameworks, public and state security monitoring and harassment, and digital surveillance.

In Chapter 9, Australia-based digital media and communications scholars Haiqing Yu (editor of the volume) and Xuanzi Xu discuss "Regulating algorithmic price discrimination on Chinese digital platforms". Based on a longitudinal study of qualitative data, this chapter examines the process involved in China's effort to regulate algorithmic price discrimination on digital consumer platforms. It pays attention to the role of social actors—consumers, consumer associations, academics, legal scholars, and state media—in the participatory process of Chinese public policymaking on algorithms regulation. It argues that regulating algorithmic price discrimination exemplifies China's dual-tracked and tiered approach to algorithm and platform governance. Algorithms governance is not just about regulating algorithmic platforms but also social governance through algorithms via platforms. Such a governance framework will reshape how the technology is built and deployed in and beyond China, impacting Chinese technology exports, the outbound platform economy, and global AI governance.

The book ends with a theoretical engagement with some of the book's key concepts in Chapter 10, "The algorithmic divide in China and an emerging comparative research agenda," by US-based legal and communication studies scholar Peter Yu. It examines the digital divide in the algorithmic context and its impact on the effective development of ADM systems. It points out that one topic that has been underexplored in ADM literature is the gap between those who have access to, or proficiency in, algorithmically enhanced or AI-driven technological products and services and those who do not. This proverbial gap resembles the digital divide, on which scholars in communication studies and other disciplines have conducted extensive research for the past three decades. Recognizing that past scholarship on the digital divide can provide helpful insights into research on the algorithmic divide, this chapter begins by identifying the similarities and differences between these two inequitable gaps. The chapter then discusses the importance of studying the algorithmic divide in China and how this study can build on, illuminate, and create synergy with China-related academic and policy research in other areas. To highlight the potential comparative insights provided by studying the algorithmic divide, the chapter concludes by examining three sets of policy responses that commentators have proposed in legal and policy literature to bridge this divide. It further contextualizes these responses in relation to local conditions in China.

To summarize, the contributors of this book sketch a variegated picture of the impacts of ADM technologies on the broader theme of digital governance and dataveillance in China. The ten chapters in this volume examine the variegated discourses and practices of automating governance in China at its theoretical, technological, sociocultural, and political dimensions. They raise questions and engage in debates on the roles of data, data ecologies, data-scoring technologies, and technological systems in automating governance in China. Taking China as a case study can play an important role in expanding and enriching the academic toolkit of terms, concepts and theories scholars can use to study the development of automated decision-making mechanisms elsewhere in the world. Existing frameworks may claim universality, but they are embedded in the political, economic and social contexts of specific locations where they originated, usually the United States or, more broadly, the West. China should neither be treated as an exception nor as a new universal model, but as a context that challenges existing frameworks.

The global dimension of China's informationization, digitization, and automation in transforming and reshaping our political systems, socioeconomic fabrics, and cultural ethos warrants nuanced, contextual, and comparative analyses from multidisciplinary researchers. By not merely contrasting Chinese to Western expectations, but considering it as a starting point in approaching questions of technology and society, this book intends to contribute to a more inclusive and culturally diverse approach to scholarship on ADM and society.

References

Backer, Larry Catá. 2019. "China's Social Credit System: Data-Driven Governance for a 'New Era.'" *Current History* 118, no. 809: 209–14. https://www.jstor.org/stable/48614454.

Bakken, Børge. 2022. *Crime and Control in China: The Myth of Harmony*. Polity Press.

Bakken, Børge. 2000. *The Exemplary Society: Human Improvement, Social Control, and the Dangers of Modernity in China*. Oxford: Oxford University Press.

Batke, Jessica and Jessica Ohlberg. 2020. "State of Surveillance". *ChinaFile*, 30 October. https://www.chinafile.com/state-surveillance-china

Central Committee and State Council. 2023. "Plan for the Overall Layout of Building a Digital China". *Xinhua*, 27 February. https://digichina.stanford.edu/work/translation-plan-for-the-overall-layout-of-building-a-digital-china/

Chen, Yixin. 2023. "Improve the social governance system (studying, promoting, and implementing the spirit of the 20th congress of the Party" [完善社会治理体系认真学习宣传贯彻党的二十大精神], Jan 11. http://theory.people.com.cn/n1/2023/0111/c40531-32604033.html

China Information Centre. 2021. "Shisiwu' xinxing jixhu sheshi jianshe zhuangjia tan [Exports discuss the construction of novel infrastructure during the '14th Five-Year Plan']". *NDRC*, 3 December. https://www.ndrc.gov.cn/wsdwhfz/202112/t20211203_1306795_ext.html

Cohen, Julie E. 2019. *Between Truth and Power: The Legal Constructions of Informational Capitalism*. Oxford University Press.

Creemers, Rogier. 2018. "China's Social Credit System: An Evolving Practice of Control". *SSRN Electronic Journal*. 10.2139/ssrn.3175792.

Creemers, Rogier. 2023. "The Great Rectification: A New Paradigm for China's Online Platform Economy". Working paper. *SSRN*. https://papers.ssrn.com/sol3/papers.cfm?abstract_id=4320952

Delury, John. 2008. "'Harmonious' in China." *Policy Review* 148: 35.

Dencik, Lina, et al. 2019. "The 'Golden View': Data-Driven Governance in the Scoring Society." *Internet Policy Review*, 8 (2). DOI: 10.14763/2019.2.1413

Flew, Terry. 2019. "The Platformized Internet: Issues for Internet Law and Policy." *Journal of Internet Law*, May, 3–16. http://dx.doi.org/10.2139/ssrn.3395901

Fourcade, Marion and Jeffrey Gordon. 2020. "Learning Like a State: Statecraft in the Digital Age." *Journal of Law and Political Economy*, 1 (1). escholarship.org/uc/item/3k16c24g

Franceschini, Ivan and Nicholas Loubere. 2022. *Global China as Method*. Cambridge University Press.

Geis, John P., and Blaine Holt 2009. "'Harmonious Society' Rise of the New China." *Strategic Studies Quarterly* 3.4: 75–94.

Hayes, Anna. 2020. "Interwoven 'Destinies': The Significance of Xinjiang to the China Dream, the Belt and Road Initiative, and the Xi Jinping Legacy." *Journal of Contemporary China* 29 (121): 31–45.

Hou, Rui, and Diana Fu. 2022. "Sorting Citizens: Governing via China's Social Credit System". *Governance* 1–XXX. doi.org/10.1111/gove.12751

Keane, Michael, et al. 2020. *China's Digital Presence in the Asia Pacific: Culture, Technology and Platforms*. London: Anthem Press. Open access: https://anthempress.com/china-s-digital-presence-in-the-asia-pacific-pdf

Knight, Adam. 2025. *Trust Is Good, Control Is Better: Technopolitical Visions and Realities in China's Social Credit System*. PhD Dissertation, Leiden University.

Kuo, Kaiser. 2024. "Political Scientist Iza Ding on Authoritarianism, Legitimacy, and 'Resilience'". *Sinica Podcast*, 25 April. https://www.sinicapodcast.com/p/political-scientist-iza-ding-on-authoritarianism

Li, Hengquang and Dahua Wu. August 19, 2022. "Four dimensions to improve digital capabilities in grassroots governance" [提升基层数字化治理能力的四个维度]. http://dangjian.people.com.cn/n1/2022/0819/c117092-32506294.html

Liang et al. 2018. "Constructing a Data-Driven Society: China's Social Credit System as a State Surveillance Infrastructure." *Policy & Internet* 10 (4): 415–453. doi.org/10.1002/poi3.183

Lyon, David. 2002. "Surveillance as Social Sorting: Computer Codes and Mobile Bodies." In *Surveillance as Social Sorting*, ed. David Lyon, 13–30. London and New York: Routledge.

Mittelstaedt, Jean Christopher. 2022. "The Grid Management System in Contemporary China: Grass-Roots Governance in Social Surveillance and Service Provision." *China Information*, 36 (1), 3–22. https://doi.org/10.1177/0920203X211011565

Mozur, Paul, Jonah Kessel, and Melissa Chan. 2019. "Made in China, Exported to the World: The Surveillance State." *New York Times*, April 24. https://www.nytimes.com/2019/04/24/technology/ecuador-surveillance-cameras-police-government.html

NDRC. 2021. "'Shisi wu' tuijin guojia zhengwu xinxihua guihua ['14th Five-Year Plan' for the Promotion of Governmental Informatization]" *Gov.cn*, 24 December. https://www.gov.cn/zhengce/zhengceku/2022-01/06/content_5666746.htm

Papagianneas, Straton. 2022. "Towards Smarter and Fairer Justice? A Review of the Chinese Scholarship on Building Smart Courts and Automating Justice." *Journal of Current Chinese Affairs* 51 (2): 327–347.

Qiu, Jack Linchuan. 2009. *Working-class Network Society: Communication Technology and the Information Have-Less in Urban China*. MIT Press.

Roehl, Ulrik B. U. 2022. "Understanding Automated Decision-Making in the Public Sector: A Classification of Automated, Administrative Decision-Making." In *Service Automation in the Public Sector: Concepts, Empirical Examples and Challenges*, eds. Gustaf Juell-Skielse, Ida Lindgren, Maria Åkesson, 35–63. Springer.

Sun, Wanning and Haiqing Yu. eds. 2022. *WeChat Diaspora: Digital Transnationalism in the Era of China's Rise*. Routledge. ISBN 9780367724276

Sigley, Gary. 2009. "Suzhi, the Body, and the Fortunes of Technoscientific Reasoning in Contemporary China," *Positions* 17 (3): 537–566.

State Council. 2015. "Outline of Operations to Stimulate the Development of Big Data." *China Copyright and Media*, 31 August. https://chinacopyrightandmedia.wordpress.com/2015/08/31/outline-of-operations-to-stimulate-the-development-of-big-data/

State Council. 2017. "New Generation Artificial Intelligence Development Plan". *DigiChina*, 1 August. https://digichina.stanford.edu/work/full-translation-chinas-new-generation-artificial-intelligence-development-plan-2017/

Tse, Jing. 2022. *Kingdom of Characters: The Language Revolution That Made China Modern*, Riverhead Book.

Tsai, Wen-Hsuan, Wang Hsin-Hsien, and Lin Ruihua. 2021. "Hobbling Big Brother: Top-Level Design and Local Discretion in China's Social Credit System." *The China Journal* 86: 1–20.

Wallace, Jeremy. 2023. *Seeking Truth and Hiding Facts: Information, Ideology, and Authoritarianism in China*. Oxford: Oxford University Press.

Wang, Maya. 2019. "Interview: China's 'Big Brother' App: Unprecedented View into Mass Surveillance of Xinjiang's Muslims." May 1. https://www.hrw.org/news/2019/05/01/interview-chinas-big-brother-app

von Glahn, Richard. 2012. "Household Registration, Property Rights, and Social Obligations in Imperial China: Principles and Practices." In: *Registration and Recognition: Documenting the Person in World History*, eds. K. Breckenridge and S. Szreter. Oxford: Oxford University Press.

Xiang, Jinglin. August 21, 2021. "Digital transformation of social governance: Orientation and trends" [社会治理数字化转型：问题指向与发展趋势]. Originally published in: *China Social Sciences Journal*. http://sass.cn/109012/64580.aspx

Yang, Shuixi, Guangxi Yang, and Caixi Liang. 2022. "Vigorously develop blockchain technology to build the 'new infrastructure' for the digital economy" [大力发展区块链技术，做好数字经济新基建]. Feb 22. *Science and Technology Daily*. http://www.news.cn/tech/20220222/d0afd5d2f57940259779201bac72d05e/c.html

Yu, Haiqing. 2019. "Social Credit: China's Automated Social Control and the Question of Choice." 13 December. https://johnmenadue.com/haiqing-yu-china-in-a-time-of-change/. Reprinted in Australia-China: A Series of Reflections, ed. Jocelyn Chey, February 2020.

Yu, Haiqing. 2022. "Living in the Era of Codes: A Reflection on China's Health Code System." *Biosocieties*. https://doi.org/10.1057/s41292-022-00290-8

Yu, Haiqing. 2023. "The Social Credit System as Method." *East Asia Forum Quarterly* 15 (1). https://www.eastasiaforum.org/2023/03/22/the-social-credit-system-as-method/

Yu, Haiqing and Jesper Zeuthen. 2023. "Local Politics in the Age of Automated Decision-Making in China: A Case Study of the Henan Health Code Scandal." *Journal of Contemporary China*. https://doi.org/10.1080/10670564.2023.2248033

Yu, Yanmin. 2008. "Corruption, Impact of Corruption, and Measures to Curtail Corruption in China." In: *China in Search of a Harmonious Society*, eds. Sujian Guo and Baogang Guo, 229–246. Boulder, CO: Rowman & Littlefield.

Zhang, Chenchen. 2020. "Governing (through) Trustworthiness: Technologies of Power and Subjectification in China's Social Credit System." *Critical Asian Studies* 52 (4): 565–588.

Zhang, Chenchen. 2023. "Hukou and Suzhi as Technologies of Governing Citizenship and Migration in China." In: *Handbook on Governmentality*, eds William Walters and Martina Tazzioli. Edward Elgar Publishing.

Zhao, Yuezhi. 2008. *Communication in China: Political Economy, Power, and Conflict*. Lanham, MD: Rowman & Littlefield.

Zhong, Jianli. 2022. "China to Introduce e-ID Cards Nationwide." *Science and Technology Daily*, March 24. http://stdaily.com/English/Feature/202203/a938465d61a14be888913f0b36541481.shtml

Zuboff, Shoshana. 2019. *The Age of Surveillance Capitalism*. London, England: Profile Books.

CHAPTER 2

Locating and Localizing Automated Decision-Making Failures in China

Xin Dai

Abstract

The chapter uses a heuristic typology to examine the diversity of real-world ADM failures, ranging from the health code debacles to credit reporting for small and medium sized enterprises, from algorithmic recommendation systems to the gig work platforms assignment system. With such examples drawn from China's experiences with deploying ADMs, the chapter cautions against the simplistic application of Western perspectives, attitudes, and discourses on ADM to the descriptive and prescriptive thinking about ADM failures in non-Western cultural and social contexts. Such a localized perspective on ADM and their failures will enable societies to better harness the decisional advantages of ADMs while living with the shortcomings.

Keywords: ADM; Bias; Noise; E-government; Culture; Discrimination

Introduction

The implementation of algorithm-driven, automated decision-making (ADM) mechanisms has become widespread globally, including certainly in China. Until recently, influential commentary on AI's public and private applications may have left people with a highly generalized impression that ADM is wreaking havoc in a typical set of problematic ways, such as by intensifying surveillance, undermining accountability, compromising due process, and magnifying systemic discrimination and unfairness (Eubanks 2018; Morozov 2011; Zuboff 2019). In light of the average citizen's ever-growing day-to-day experiences with ADM, however, it should have become increasingly clear to many that, as each application of ADM often affects different segments of the population in distinct ways, the development poses a multitude of problems and challenges in response to which we require much more than a universalist understanding and generic prescriptions.

This chapter aims to shed further light on the diversity of real-world ADM applications and, more specifically, the varying nature of their failures that cause harm and controversies. For a working definition, the term "ADM failures" in this chapter refers to the discrepancy between the expected and actual effects of an ADM application, a discrepancy presumed to be a primary source of public discontent and frustration. By viewing ADM failures through such a lens, this chapter offers a pragmatic approach to understanding ADM's underlying issues in a manner that also goes beyond merely technical performance evaluations. In particular, as I focus on analyzing how ADM mechanisms fail in terms of not aligning satisfactorily with localized societal expectations, consciously or subconsciously held by interested parties, I argue that it is crucial to identify and understand such expectations as baselines when assessing ADM's performance. Through such analysis, this chapter aims to reorient law and policy deliberations on ADM towards an analysis that pays greater attention to localized preferences and issue-specific remedies, and to the discourse of universal values.

Drawing on the work of Kahneman, Sibony and Sunstein (2021) on the psychology of decision-making, this chapter sets forth a heuristic typology in the form of a two-by-two matrix as a heuristic device which allows for the presentation of four broad types of ADM failure along two fundamental dimensions. The first dimension is "bias," which describes the extent to which decision outcomes produced by an ADM mechanism deviate systematically from its intended optimization goal. The second dimension is "noise," referring to the level of variability in outcomes generated by an ADM, with high noise rendering the mechanism unable to produce reliably predictable decisions. By organizing the discussion within such a framework, the chapter examines various real-world ADM applications in the contemporary Chinese context, demonstrating how ADM failures in these varying scenarios represent distinct challenges and necessitate different solutions.

This approach is useful not only for locating the problematic features of each ADM, but also for revealing the localized nature of societal concerns over ADM given that societies are not uniformly affected by the same types of ADM failure. As explained later, the "high bias, low noise" type of ADM failure, for example, illustrated by discriminatory algorithms that are accused of systematically disadvantaging already disenfranchised social groups such as racial minorities, has received mounting attention in Western literature. Such ADM failures are of much less concern in the more ethnically homogeneous Chinese society where discriminatory treatment operates and is debated along very different lines. To develop effective and pragmatic policy responses that will allow us to harness the decisional advantages of ADM as intended (Kleinberg et al. 2018; Sunstein 2022) and also to prevent and mitigate excessive harm, it is essential to acquire a contextualized and culturally informed understanding of ADM failures.

This chapter has three sections. Section one introduces the general approach this chapter takes in differentiating ADM failures. Section two details the four types of ADM failures, with examples from Chinese practices that illustrate the need for different policy understanding and prescriptions. Section three takes further stock of key lessons from the chapter's analysis before the chapter concludes.

Differentiating ADM Failures

To illustrate why a more contextualized understanding of ADM failures is important, let's first consider two Chinese consumers who complain, both in high-profile court cases, about mistreatment by platform algorithms. The first complains that the search engine platform uses cookies to track terms she once used for a search, including "weight loss", "abortion", and "breast implant", which results in this consumer seeing offensive advertisements in relation to those terms as she views other websites.[4] The second consumer complains about seeing higher prices for hotel rooms compared to other users. That is because the algorithm used by her online travel agency (OTA) platform correctly identifies her as a long-time, loyal user who is willing to pay more and is less sensitive to price than the not-so-loyal customer.[5]

Both cases illustrate the use of problematic ADM mechanisms by commercial platform companies that potentially violate consumers' rights, interests, or legitimate expectations of fair treatment, but it takes closer examination to appreciate that the ADM system in the two cases causes problems that differ. In the first case, the automated system for advertisement delivery can be considered an ADM failure, not only because it potentially offends consumers but also because harming user experience ultimately goes against the interests of the firms using the algorithm, that is, firms seek to improve advertisement efficacy through a greater level of personalized access. In other words, such an ADM process would fail to perform the function desired by those that deploy it. By contrast, in the second case the algorithm in question is doing exactly what the platform operator intended it to do—personalizing pricing terms based on each consumer's willingness to pay. It is considered a "failure" only from the perspective of the consumer as opposed to that of the profit-maximizing firm.

For policy and regulatory discussions, it is important to first understand the varied nature of problematic ADM failures. In the two cases of algorithmic consumer violations discussed above, for example, one must see that they involve

[4] The fact pattern of this hypothetical is adapted from one of China's earliest cookie litigations involving the search giant Baidu. See "Zhu Ye yu Beijing Baidu Wang Xun Keji Gongsi Yinsi Quan Jiufen Shangsu An"朱烨与北京百度网讯科技公司隐私权纠纷上诉案[The Appellate Case of Zhu Ye versus Baidu on Privacy Right Disputes]. Jiangsu Sheng Nanjing Shi Zhongji Renmin Fayuan 江苏省南京市中级人民法院 [Intermediate Court of Nanjing, Jiangsu](2015).

[5] The fact pattern of this hypothetical is adapted from a well-reported case in China about big data price discrimination in recent years involving leading OTA platform Ctrip.com. See "Shanghai Xiecheng Shangwu Youxian Gongsi yu Hu Mou Qinquan Zeren Jiufen Shangsu An"上海携程商务有限公司与胡某侵权责任纠纷上诉案[The Appellate Case of Ctrip.com versus Hu on Tort Liability Disputes]. Zhejiang Sheng Shaoxing Shi Zhongji Renmin Fayuan 浙江省绍兴市中级人民法院 [Intermediate Court of Shaoxing, Zhejiang] (2021).

the use of ADM mechanisms that are problematic in different ways. Since the controllers of these ADM mechanisms are not similarly motivated to address their purported failures, generic regulatory and policy approaches aimed at strengthening consumer protection are either unnecessary, in the one case, or incapable of effectively changing behavior in the other different context.[6]

To better organize the discussion about how problems associated with ADM may be differentiated, this chapter adopts a simple heuristic typology as described below. To be sure, there are many other types of classification potentially available for the same purpose. In the practical realm of law and policy, for example, regulatory authorities in China and Europe have already developed formal classification systems for ADM along such dimensions as functionality, societal impact, levels of risks, and so on (also see Chapter 9 in this book).[7] This chapter does not intend to directly contribute to the further refinement of these existing classifications; instead, it aims to offer an alternative lens for considering why and how each ADM fails to deliver the expected results. Such an exercise may also shed light on potential directions for prescriptive deliberations.

Kahneman et al.'s Two-by-Two Matrix

This chapter adopts a heuristic framework that is adapted from a simple two-by-two matrix typology for decision-making quality. In their book *Noise*, Kahneman, Sunstein and Sibony propose that the effectiveness of decision-making activities,

[6] For example, it is probably sufficient for regulators to require the search engine platform to ensure users receive notice about their cookie use, because the users would not be able to make informed choices about whether to accept the cookie options. They will know they are being tracked easily upon seeing the offensive personalized ads and may complain to the firm which may adjust its practices in response to user experience. It is unlikely to be adequate, in the algorithmic price discrimination case, that the platform is merely required to give notice to users about such practices in the form of user agreements, which users typically ignore and fail to read before giving consent. That is because algorithmic price discrimination is mostly difficult for average users to identify.

[7] The Chinese regulatory authority has long adopted the over-arching principle of "differentiated regulation and management with grading and classification" (分级分类监管) for the overall regulation of the digital economy. In 2021, the Cyberspace Administration of China, a key Internet regulator, issued a highly anticipated regulation on recommendation algorithms (hereinafter "Recommendation Algorithms Rules") that seeks to establish a classification system for algorithms based on a range of factors, such as the algorithms' influence on public opinion or social mobilization, the content they push, the size of their user base, the importance of the data processed, the extent of their intervention in user behavior, and other relevant considerations. See Hu Lian Wang Xinxi Fuwu Suanfa Tuijian Guanli Guiding (互联网信息服务算法推荐管理规定) [Provisions on the Administration of Algorithm-generated Recommendations for Internet Information Services] (promulgated by Cyberspace Administration of China et al. on December 31, 2021).

whether performed by humans or machines, can be evaluated along two dimensions: bias and noise (Kahneman, Sibony and Sunstein 2021, 3). The level of bias measures the extent to which decisions deviate systematically from the intended optimization target. While the term "bias" is often associated with socially undesirable discriminatory practices, in the context of decision science it refers to the factual, objectively measurable deviations from the optimal outcome, without necessarily implying the presence of intentional discrimination by the decision-maker. The level of "noise", meanwhile, measures how scattered the outcomes are when generated by a decision-making tool or mechanism. Excessive noise indicates unpredictability and unreliability, resulting in unwanted variability of judgments (Sunstein 2022).

Using these two dimensions, Kahneman et al. constructed a simple two-by-two matrix that places decision-making procedures into four broad categories, as illustrated in Figure 2.1, using the metaphor of an archery contest.

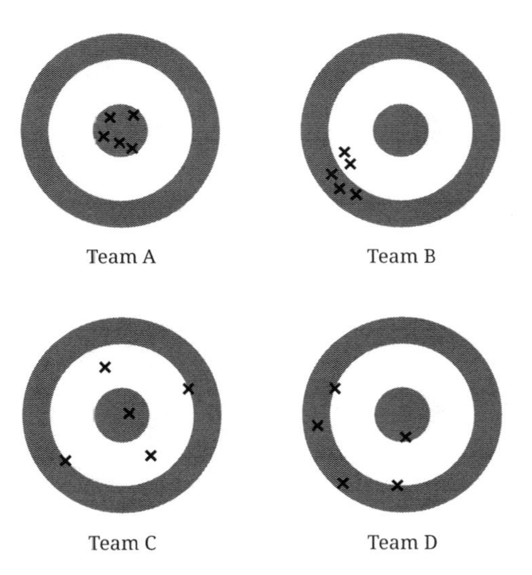

Figure 2.1 Kahneman et al.'s (2021, 3) archery contest matrix

The upper-left Team A, hitting the bullseye reliably, represents effective decision-making with minimal bias and noise. The upper-right Team B represents biased but not noisy decision-making, as their arrow shots systematically deviate from the bullseye but consistently fall in places close to each other. The lower-left Team C represents noisy but not biased decision-making, as the outcomes are widely scattered without systematic bias in any direction. The lower-right Team D, shooting arrows that fall on spots systematically biased towards the lower-left and also scattered widely, represents decision-making that is both biased and noisy.

It is important to note that Kahneman et al.'s analysis is not specifically about machine-based ADM; instead, it applies to decision-making in general. Indeed, what they intend to highlight is that human decision-making in various high-stakes contexts, such as court adjudication and public policy, is prone to significant problems of bias, noise, or both (Kahneman, Sibony and Sunstein 2021). Contrary to the prevailing discourse against machine decisions, they argue in a somewhat contrarian manner that computer algorithms should be preferred over human decision-makers in many contexts where humans are susceptible to serious shortcomings in terms of bias and noise (Kahneman, Sibony and Sunstein 2021; Sunstein 2022).

The Heuristic Framework for ADM Failures

This chapter takes Kahneman et al.'s approach one step further. While computer algorithms indeed offer some effective solutions to the various inadequacies in human decision-making capabilities, it is also well-known that all machines, or ADM mechanisms, are not created equal in terms of being useful and reliable. Critics rightly point out that real-world ADM mechanisms can also be biased and/or noisy. That is especially true when they are assessed against the societal expectations underlying their deployment, expectations that demand higher quality performance from an ADM mechanism than the human decision-makers they replace or supplement.

In other words, ADM from time to time does fail to meet human expectations. As this chapter focuses on considering ADM failures in such a context, it proposes to capture further differences among ADM failures with the following two-by-two matrix framework, adapted from Kahenman et al..

Table 2.1 The matrix of ADM failures (source: author)

		Bias	
		High	**Low**
Noise	**High**	Health code debacles in 2022 Platform content moderation systems	Crude e-government systems Offensive algorithmic recommendations
	Low	Credit reporting for SMEs Resume filtering and background check systems	Gig work platforms' assignment system Algorithmic price discrimination

Under this matrix framework (Table 2.1), each of the four quadrants represents one type of ADM failure that I consider different from others both in their nature and in terms of appropriate responses. The upper-left quadrant represents "high bias, high noise" ADM failures, in which the mechanism consistently produces partial, unreliable, and unstable outcomes, resulting in both inequity and chaos. One notable example in the Chinese context of such an ADM failure is that of the "Health Code" systems[8] (Yu 2022) used for the Chinese government's pandemic control policy, in particular in the second half of 2022, prior to China's exit from the "dynamic zero-covid" policy.[9] Another less dramatic but more persistent example of such ADM failure concerns the content moderation or censorship systems used by China's social media platforms, long complained about because of both biased and inexplicable decisions in blocking user content.[10] Both examples, in particular their implications for our thinking about how to respond to ADM failures, will be discussed further in the next section.

The lower-left quadrant represents "high bias, low noise" ADM failures, where the mechanisms produce decision outcomes that are systematically biased in favor of certain decision targets at the expense of others, and where such patterns of bias are either unintentional or without official endorsement from the entity that controls and deploys such ADM. In the Western discourse, these failures, often generally labelled as "algorithmic discrimination," are a significant concern in relation to algorithm applications in high-stakes decision-making scenarios ranging from criminal justice and credit reporting to school admissions. In China, by contrast, this type of ADM failure is in fact much less salient in public discourse; for example, despite serious misunderstandings about China's social credit system in the West (Pence 2018), the ADM mechanisms incorporated in that project have no conceivable or perceived effects that disadvantage minority ethnic groups. One Chinese illustration of the "high bias, low noise" ADM failures concerning bias against small-medium enterprises (SMEs) is in commercial credit reporting.

[8] For simplification purposes, "health code" in this chapter refers to both health code (健康码) and itinerary code (行程码) systems. The former is a provincial system whereas the latter is a national system.

[9] See e.g. "Hui Bu Qu de Jia, Shang Bu Liao de Ban: Zai Beijing Gongzuo de Langfang Ren" ("回不去的家，上不了的班"：在北京工作的廊坊人) [Not Going Home, Not Going to Work: Langfang Residents Working in Beijing], San Lian Shenghuo Zhoukan (三联生活周刊)[San Lian Weekly], Jun. 20, 2022, http://ny.zdline.cn/mobile/audioText/?artId=166998&sm=app&parentUserId=1645858, last visited on Oct.4, 2023.

[10] There are many complaints from users about not understanding the reasons behind their submitted content being blocked on social media sites. See e.g. Weixin Kaifang Shequ (微信开放社区)[Weixin's Community Board], https://developers.weixin.qq.com/community/minihome/doc/0000e63e1acbb8b5e2be38a895b000?blockType=99, last visited on Oct.4, 2023. There are also many "tutorials" for avoiding sensitive areas. See e.g. "Wei Shenme Wenzhang Zong Shi Shenhe Bu Guo? Zhe Si Ge Leiqu Ni Bi Kai le ma?"(为什么文章总是审核不过?这4个雷区你避开了吗?)[Why do your posts always fail to pass the review? Have you circumscribed these four forbidden zones?], https://baijiahao.baidu.com/s?id=1726987746125346559&wfr=spider&for=pc, last visited on Oct.4, 2023.

Resumé screening systems and other employment-related ADM mechanisms are also widely agonized over in China due to the risk of failure due to bias, such as against female job seekers, although reported and systemically verified cases cannot be accessed in China at this point.

The upper-right quadrant represents "low bias, high noise" ADM failures, where outcomes are considered unpredictable and erroneous, though without systemic bias. The Chinese examples that will be further discussed in the next section include occasions of ineffective and flawed e-government systems. Most of these failures may sound quite banal, but still matter a great deal to citizens; for example, there are reported cases of the ADM system mistakenly terminating people's pension payments due to system and data analysis errors.[11] Often less associated with the problem of noise are the recommendation algorithms used by platform companies such as social media firms which personalize the delivery of content to users. Yet, as the previously mentioned Baidu case suggests, often these ADM systems can be considered as failures for operating with a level of noise—in terms of their delivering too much obnoxious content to users, content not considered tolerable.

Finally, the lower-right quadrant represents "low bias, low noise" ADM failures. This may sound counter-intuitive because "low bias, low noise" *should* mean that the ADM systems are hitting their intended target reliably. And indeed this is where this chapter's typological framework diverges from that of Kahneman et al. For the purpose of this chapter, this category of ADM failure is of particular importance, as it clearly presents the critical baseline or reference point question in analyzing ADM, that is, societal expectations of ADM performance, as opposed to merely what the controller and deployer of the ADM intend it to do. Specifically, those ADM systems that fail in this sense attract protest mostly because the goal the system tries to optimize clashes with other social values. Revealing examples in this category will be found in the outcries against gig-work platforms' algorithmic work assignment systems. Debates over so-called algorithmic price discrimination as practiced by e-commerce platforms also shed light on why these ADM failures may be most difficult for policymakers to properly address.

[11] "Yisi You Dian Siwang Ting Fa Yanglao Jin shi Dianxing de Lanzheng" ("疑似有点死亡"停发养老金, 是典型的懒政)[It is classic lazy policy to automatically terminate one's retirement benefits when the system mistakenly identifies recipient as "possibly deceased"], https://baijiahao.baidu.com/s?id=1665210163731713189&wfr=spider&for=pc, last visited on Oct.4, 2023.

Locating and Localizing ADM Failures with the Two-by-Two Matrix Framework

As preliminary to the discussion below, a couple of caveats are in order. First, the two-by-two typology is obviously crude in its categorization; more granular classifications with additional levels of bias or noise (such as not only "high," "low," but also "medium") are certainly possible and analytically precise; however, the construction is intended here for its virtue in not only being simple but also easier to implement in practice. Second, the examples used here are for illustrative purposes, and I realize that interpretations of the nature of the failure of each ADM may vary. Different decisions about the appropriate category of one ADM failure under the framework may all be valid as long as these determinations are made consistently for the purpose of the same analysis. Nevertheless, if a significant number of borderline scenarios emerged that caused people to struggle with sorting them into one, not multiple, quadrants, then the framework's usefulness would certainly be questioned.

High bias, high noise ADM failures

At first sight, the high bias, high noise ADM failures concern situations where the ADM system is obviously not up to the job. It seems intuitively tempting to conclude, therefore, that the right response is to just discard such ADM systems, and also to make the effort to avoid adopting similarly flawed mechanisms.

The reality confronting society is often more complex, however. When we consider an ADM as "high bias, high noise", it is frequently not being compared only to human decision-makers; in those comparisons we conclude that the ADM does the job even less effectively than humans; instead, even where the ADM, despite its flaws, outperforms the corresponding human decision-makers in terms of speed, scale, and accuracy, it may still fall short of our idealized notion of a high-performance ADM, and therefore be deemed as a failure. This is particularly likely in high-stakes decision-making scenarios where the general public's tolerance for the level of bias and noise is particularly low. One good example is law and justice. The fact that ADM may significantly but far from perfectly reduce both bias and noise in important legal decisions, such as pre-trial decisions for granting bail, is highly unlikely to persuade people to accept the use of machines instead of human judges (Kleinberg et al. 2018).

For useful ADM systems that nevertheless fail in the sense they do not meet people's high expectations, the rational reaction is not simply to discard them. What therefore should be done with them? One approach is to treat them as largely products of utility but with known and unknown defects, and regulate them accordingly. To the extent we choose to allow their use, we should make an effort to construct a proper institutional structure that produces incentives for those entities designing, developing, and using these ADM mechanisms so that, under pressure from both competition and the prospect of liability for harm, such entities will be induced to invest in refining the algorithms to continually reduce bias and noise.

The evolution of the ADM systems used by Chinese social media platforms for their content moderation practices, in particular the pre-screening systems that review user-generated content before they can be shown to the public, serves as an illustration of how firms may be motivated to constantly refine a high bias and high noise system under both competitive and regulatory pressures. The content moderation systems, as of today, remain the target of considerable societal dissatisfaction for their apparent bias against some but not all types of offensive or sensitive content, and for the obvious presence of noise, with users often finding that the blocking decisions against certain content they submit to the sites are inexplicable.[12] Nonetheless, in light of the very crude keyword filtering and blocking systems with which censors started out in the late 1990s, the pressures social media operators face to balance the need for attracting and retaining user flows, on the one hand, and complying with content rules on the other, have led to the development of increasingly sophisticated and efficient systems,[13] especially taking into account the fact that the task of content moderation has also become ever more challenging over time with changes in both the technology and the censorship system.

One important thing to keep in mind, however, is that the presence of adequate competitive pressures cannot be taken for granted. The digital economy, for example, is well known for hosting problematic market structures, not particularly conducive to competition (Tirole 2017, 392–400), while for ADM systems deployed in the public sector, competition among system providers is also less dependable as a means of ensuring product improvement. The necessity for effective regulatory intervention, therefore, is unequivocal in addressing ADM failures.

But what kind of intervention is necessary and desirable? A comprehensive survey of considerations associated with regulating artificial intelligence, with AI readily available nowadays (Chesterman 2021), is certainly not within the intended scope of this chapter. Of note here is that one commonly proposed effective response to ADM failure—but one unlikely to be implemented—is to increase the availability of human intervention. In other words, to the extent that an ADM with bias and noise higher than socially tolerable levels will remain in use, it is often advised that humans must always stay "in the loop" to supervise and intervene in the machine-made decisions (McKendrick and Thurai 2022). This proposal is

[12] See supra note 8

[13] For an example of the current systems, see the description in a media company's promotional material in Lu et al. (2022).

sound and appealing in the abstract, but the potential risks of introducing human intervention to the process is under-appreciated.

A notable example in the Chinese context is the health code systems' collapse that occurred in the second half of 2022. The health code systems were implemented in China's zero-covid policy era, that is, from early 2020 (Yu 2022). As previously mentioned, during those final months, before China's exit from its zero-covid policy, many individuals across the country faced erratic travel restrictions because the pandemic control surveillance system assigned them "red codes" or "warning windows," signifying the assessment of a high exposure risk. People were frustrated by the bias and noise exhibited in the machine decisions. The restrictions were blanket measures, highly biased against all individuals with travel histories in typically broad geographical areas designated as high risk. The result was that many low exposure risk individuals who happened to have the "wrong" travel history were often subject to restrictions. There were also frequent errors in which individuals without travel history to those risk areas were incorrectly flagged, and vice versa.[14] Initially, the blame was more often directed towards the perceived inferiority of the technologies underlying the health code systems; however, it was eventually realized that the determination of the code color involved significant human participation, thus introducing opportunities for considerable tampering and manipulation. Government officials in Henan province, for instance, on one occasion assigned red codes to all residents of a particular county to prevent them from leaving home to protest against a large-scale banking fraud scheme.[15] In Beijing it was street-level bureaucrats and community workers, as opposed to computers, who were in the end tasked with reviewing and approving petitions for code restoration for people struggling to return home. These manual interventions led to considerable confusion and seemingly arbitrary decisions across communities.[16]

[14] "Mei Quguo Zhong Gao Fengxian Diqu, Xingcheng Ma ye Dai Xinghao? Xingcheng Ma Lao Shi Chucuo Gai Zenme Ban?" (没去过中高风险地区, 行程码也带星号?行程码老是出错该怎么办?)[Why does the itin code carry an asterisk for someone without travel history to areas of high and intermediate levels of risk? What should you do if the itin code keeps getting things wrong?], https://baijiahao.baidu.com/s?id=1731798304688237415&wfr=spider&for=pc, last visited on Oct.4, 2023.

[15] "Quan Xian Jumin Bei Fu Hong Huang Ma, You Bei Kexue Jingzhun Fang Kong Jingshen"(全县居民被赋红黄码, 有悖科学精准防控精神)[It is against the spirit of scientific and precise pandemic control to give red and yellow codes to all residents of the entire county], https://baijiahao.baidu.com/s?id=1740203320319342383&wfr=spider&for=pc, last visited on Oct.4, 2023.

[16] Beijing's health code system had a unique feature: the screen of a person's code could freeze when it showed a message window requiring a person to get a PCR test as soon as possible—or they would not have a valid code for entering public places. Such a person's code could only be restored after they submitted PCR test results and other information to an electronic system, but the review was often manually performed by community workers in the back office. One may still find many accounts on social media sites of ordinary citizens' personal frustrations when dealing with such a

The health code debacle reveals, therefore, a clear lesson: human intervention is not necessarily the right solution to high bias, high noise ADM failures.

High bias, low noise ADM failures

The second category of ADM failures pertains to situations where the primary concern is the presence of systematic bias in the algorithmic decision-making. It is important to note that this does not imply such ADM mechanisms are completely free of noise-related issues; it simply means that the predominant concern is bias and the resulting discrimination implications.

Algorithmic discrimination and its relationship to background societal discrimination have been extensively discussed. While it may be unnecessary to reiterate conventional wisdom, for what it is worth, one primary line of critique against ADM is that it reinforces or amplifies existing social discrimination (Ajunwa 2020). In the Chinese context, however, it is important to note that discrimination does not become manifest in the same way as it does in the West, where the foremost concern is discrimination based on race and ethnicity. Societal discrimination in China, by contrast, does not gravitate towards one singular, salient dimension. It is exhibited instead in various forms and in multiple dimensions such as gender, age, physical disability, Hepatitis B carrier status, height, educational background, urban residential status, home region, and other dimensions (Zeng 2007). These distinctions, problematic as they are, may underlie decisions taken in formal contexts such as public and private sector employment, as well as in informal contexts such as marriage and business associations. Without a single focal point like race, however, social discrimination in China is more diverse and less entrenched; consequently, discriminatory ADM has not generated the same level of anxiety as in the West. On the practical side, even scholars writing about the subject would admit that real-world cases such as discriminatory resumé-filtering algorithms against females are "likely to emerge" but they have not been identified as yet (Yan 2021).

A real case of high bias, low noise ADM failure in China may be found in the context of credit reporting. As previously noted, it is certainly more banal than is suggested by the popular myth regarding the social credit scoring of individual citizens. China's nationwide business credit reporting system, premised on a centralized data system at the People's Bank of China (PBOC), the nation's central bank, has long been known to systematically favor established large-size enterprises,

system. See e.g. "Ren Buzai Beijing Tanchuang San Yijing Shensu le + Haishi Shenhe Wei Tongguo"(人不在北京弹窗3已经申诉了+还是审核未通过)[Not being physically in Beijing and having appealed the Type 3 push message but still no approval], https://wen.baidu.com/question/505282648415524604.html, last visited on Oct.4, 2023.

especially state-controlled enterprises, over small and medium-sized enterprises (SME) in determining access to credit. Such decision patterns are considered biased, for the purpose of this chapter, not only because the results preference one group of decision targets. More importantly, it should be realized that the biased pattern is not entirely intentional; rather, it results largely from the absence of credit data for smaller firms in the PBOC's data system. Such biased credit assessment ADM fails therefore in the sense that its designed purpose, unfulfilled, is to do a better job in terms of optimizing access to credit.

Irrespective of whether the high bias ADM failures appear in "exciting" or banal scenarios, I contend that there are fundamentally no new challenges they pose when compared to the discriminatory practices of the pre-ADM era. The most relevant response remains the established legal or regulatory safeguards, such as *ex ante* exclusion of sensitive data, the right to rejection of ADM, and *ex post* rights to explanation and petition that aim to police data processing practices considered prone to high risk of discrimination. Those regulatory approaches are far from perfect solutions, but they could offer meaningful mitigations, at least in certain contexts. More importantly, contrary to the popular notion of ADM being an impenetrable "black box," the discriminatory effect of ADM is often more detectable than that arising from human-made decisions; the door is therefore open to innovative and more effective remedies, such as so-called "big data affirmative action" (Salib 2022). That is particularly true where the controller of the ADM is motivated to reduce unintended bias. In the case of credit assessment for SME, for example, when the monetary policy calls for credit expansion rather than contraction, not only will bias against SME be uncovered through audits, but the credit evaluation algorithm will be adjusted to significantly reduce such bias in the future.[17]

Low bias, high noise ADM failures

The third category of ADM failure involves situations where an ADM mechanism produces inexplicable errors or exacerbates the difficulties and frustrations people face when performing certain tasks, rather than reducing such annoyances.

In the Chinese context, the e-government systems adopted in many localities, notable for their crude, rudimentary, and inadequate design and implementation, fall into this category. While there are cities such as Shanghai, Hangzhou, and Shenzhen where governments have successfully implemented and operated sophisticated and functional systems, elsewhere it is common that citizens

[17] Key facts on SME financing in People's Republic of China, Financing SMEs and Entrepreneurs 2020: An OECD Scoreboard, https://www.oecd-ilibrary.org/sites/31f5c0a1-en/index.html?itemId=/content/component/31f5c0a1-en, last visited on Oct.4, 2023.

complain about, as opposed to welcome, the e-government rollouts that purport to make access to government services much more convenient and efficient for them. A considerable number of the issues concern incomprehensible errors that people encounter when forced to use e-government systems to manage their lives. These errors may be attributed to underlying data quality issues or system configuration problems (Zheng et al. 2021). Ironically, in many localities where the lobbies in government agency buildings contain physical kiosks for automated processing of visitors' routine matters, designated staff members are also often assigned to assist people in using the machines and, more importantly, in directing them to again wait in line before human-staffed counters, since the machines are expected to frequently fail.[18]

While such inconvenience may seem tolerable or even trivial at an individual level, the cumulative inefficiency resulting from these high noise ADM failures can still be significant. In principle, government authorities can ensure continual improvement by investing in upgrading data, infrastructure, and system design. As mentioned earlier, one needs to remember that there are relatively more successful examples where e-government initiatives have indeed improved citizens' experiences and lowered the overall cost of bureaucracy to society. As discussed, when it comes to public sector ADM applications, however, it is not surprising to find that the motivation for meaningful improvement is far from universal or consistent. As extensively discussed in administrative law literature, the most popular proposal in response to low bias, high noise ADM failures, especially in the public sector, is to implement a range of procedural safeguards so that individuals are afforded opportunities to identify, contest, and correct mistakes made by the machines (Crawford and Schultz 2014; Kaminski and Urban 2021).

Of course, low bias, high noise ADM failures are also observed in market settings. One conceptually interesting example I think worth discussing here is the personalized recommendation systems used by commercial internet platforms, such as e-commerce and social media sites. While it is common to hear complaints about bias in these algorithms, it is interesting to note that they are also criticized, although not in such terminology, for being noisy, as they fail from time to time to recommend or deliver content to users according to their preferences. In some markets, for example, there may be a statistically valid correlation between people searching for milk formula and their interest in diapers; in other markets, however, where the correlation exists but is weak, the margin of error may be large enough so that a good many consumers will be offended when they receive recommendations for diapers simply because they searched for formula. In other words, the

[18] For examples of popular complaints, one may search on the popular message board Tieba, such as at https://tieba.baidu.com/p/6662877323, last visited on Oct.4, 2023

level of noise can have a negative impact on consumer experience. Social media recommendation algorithms face even more severe problems in this regard when a single search for potentially controversial news inundates a user with a non-stop slew of inappropriate content. Such an "in your face" style of personalization is often not consistent with the platform's commercial interest either.

Presumably therefore there would be market pressure for firms to address the loss of consumers due to the high noise in these scenarios. The precondition for this, however, is the presence of market competition. Through regulation such as the Recommendation Algorithms Rules, authorities have attempted to exert regulatory pressure on platforms using these ADM systems to improve consumer experience. Nonetheless, one challenging aspect is that when regulators intervene to reduce noise, they may inadvertently introduce more bias; for instance, while the Recommendation Algorithms Rules require the platform algorithms to reduce the visibility of unwanted and inappropriate messages, they also require that official propaganda be given priority in recommendations, thereby creating high risk of bias.

Low bias, low noise ADM failures

As previously noted, this chapter's typological framework diverges from that of Kahneman et al. most notably by considering certain "low bias, low noise" ADM systems as "failures." As noted, this may seem counterintuitive because, while these algorithmic systems appear to optimize their intended decision targets, they are considered failures here due to the disconnection between the desired optimization target of the ADM user and desirable societal goals. While the developers and users of these ADM systems may be satisfied with their performance, or at least find the level of noise and bias tolerable, broader society, or a significant segment of it, may object to the decisions made by these systems.

One commonly discussed example of this type of ADM failure is the work assignment and monitoring system used by gig-work platforms, such as food delivery platforms. Their algorithms match available workers with consumer task requests and dictate delivery routes to optimize efficiency and consumer satisfaction. Generally, this mechanism contributes to high consumer satisfaction and platform profits; however, in China they have also sparked considerable resentment among gig workers, such as "food delivery riders," who have voiced their grievances through investigative journalist reportage.[19] These workers protest that the system places excessive demands on them, neglecting their need for safety, leisure time, and dignity.

[19] See, Lai, Youxuan赖祐萱. "Waimai Qishou, Kun zai Xitong li" (外卖骑手, 困在系统里)[Delivery riders stuck in the system], https://baijiahao.baidu.com/s?id=1677231323622016633&wfr=spider&for=pc, last visited on Oct.4, 2023.

Another example concerns the pricing algorithms used by e-commerce platforms. Although the term "price discrimination" suggests high bias, from the platform's perspective the pricing ADM is actually low bias because its results align with the intended optimization target which is to price products based on consumers' willingness to pay, maximizing the platform's profit. Unsurprisingly, consumers and society at large have engaged in intense debates regarding the fairness and efficiency implications of these ADM systems (see the chapter by Yu and Xu in this volume).

The low bias, low noise ADM failures pose different challenges for policymakers compared to the previous three types in which efforts were primarily focused on reducing bias, noise, or both. The key question here is how society can reconcile the conflict between different values or claims for priority among different interest groups. In the case of gig-work platform ADM systems, for example, in the underlying conflict between consumers and workers the Chinese government has taken an unequivocal stance in support of workers, having issued a range of policy directives requiring platforms to improve working conditions.[20] Facing changed economic realities, however, especially during the post-Covid recession, it is certainly not unthinkable that government authorities may step back from more pro-labor policies, now more concerned about overall employment rates. In relation to price discrimination, the question becomes even more complex, delving as it does into the heart of microeconomic theory. Does the particular type of price discrimination practiced by the platform increase or decrease general welfare? If it creates a larger pie, should society forgo it and instead have the transaction parties share a smaller pie, just in a more equitable manner? The answers to both questions, as well known, are highly contingent (Amstrong 2013). It has consequently been quite difficult to formulate policy responses to these contested failures.

What this suggests, therefore, is that the challenges posed by the low bias, low noise ADM failures arise not solely because of the application of ADM, but also because the underlying questions are fundamental ones that humans have grappled with for a long time without definitive solutions. In these situations, one commonly proposed approach that society may adopt is to call for the introduction of accountability regimes for those entities, public or private, which use these ADM systems (Kroll et al. 2017). It is important to acknowledge, however, that these accountability regimes by themselves simply do not answer the underlying questions directly. Instead, what they can do is ensure that the ADM mechanism's optimization target aligns with society's overall preferences, that is, the public

[20] The Recommendation Algorithms Rules, for example, include a provision (Section 20) that specifically requires that gig work platforms, in using algorithms to assign tasks to workers, must protect their rights to remuneration and rest.

interest. Unfortunately, public preferences or the public interest itself can be quite volatile and subject to change in relation to such matters. One potential danger is that, in the process of demanding accountability, excessive human intervention in the system may again be introduced, which will transform, as opposed to solve, the low bias, low noise ADM failures into high bias, high noise failures.

Further Discussion

The above discussion has demonstrated that ADM failures vary across contexts such that social and policy responses to them should also vary. In the following, I draw attention to the three most noteworthy points on which the foregoing discussion sheds some light.

First, despite often exaggerated hype in the popular discourse about the risk of social harm from ADM, it is important to recognize that most ADM systems that people encounter in their daily lives are tools developed in response to societal demand for improving decision-making previously done by humans. As a result, to assess whether and how seriously they fail, it is advisable that we use realistic and pragmatic benchmarks. As with any other products in the market, it should be expected that there are going to be many cases of product defects, which means that an ADM system may plainly fail to improve decision-making quality along bias, noise, or both dimensions—just like the human decision-makers previously in place. In these cases, society may realistically consider such ADM failures very much as product defects to be addressed through the known market and through institutional mechanisms available in the conventional regulatory toolbox. In particular, in contexts where market incentives are inadequate, such as in public sector ADM applications, stronger regulatory oversight and intervention could be necessary.

Second, the previous analysis reveals that ADM failures are not solely an objective or technical phenomenon but are instead cultural constructs that are shaped by subjective societal expectations. According to some recent accounts, China remains one of several major countries (among other developing countries such as Brazil, India, and South Africa) to hold more positive attitudes towards AI than Western countries (KPMG & University of Queensland 2023). That may plausibly lead the Chinese public to show greater tolerance for ADM failures that are seen as technologically improvable and therefore temporary. In particular, public-service ADM systems managed by the central government can be perceived as more neutral, reliable, and efficient when compared to local government systems and street-level bureaucrats who are often considered arbitrary, incompetent, and untrustworthy (Chen 2017). This would challenge the over-arching Western presumption that ADM would fundamentally threaten, as opposed to reinforce, the legitimacy of public institutions (Calo and Citron 2021). These patterns of cultural perception and attitudes may affect approaches taken in response to ADM failures, such as whether greater tolerance and patience will be afforded, allowing these systems the privilege of trial and error. In particular, for those ADM systems already achieving improvement over human decisions but still falling short of some subjectively high expectations, the cultural and discursive environment in which they operate can often be determinative in their eventual trajectory.

Third, one cautionary tale that emerges from examining the Chinese practices is that there are potential dangers of excessive human interference in response to ADM failures. The prevailing discourse tends to emphasize risks and dangers associated with fully automated systems and advocates for human-in-the-loop approaches. Involving humans in algorithm-driven decision-making processes is both sensible and necessary in principle, but it is also critical that the nature and degree of human involvement be carefully deliberated and implemented. An overemphasis on human involvement as a response to ADM failures may backfire and lead to unintended consequences, such as introducing higher levels of bias and noise. Moreover, as illustrated in the Chinese cases, such as that of the failure of the health code systems (see, for example, Yu and Zeuthen 2023), it is important to note that less than well calibrated human intervention may significantly undermine societal support for the use of ADM systems in general. What may be at work is the social psychological mechanism of "betrayal aversion," which refers to the stronger negative emotional reaction of people to bad outcomes in scenarios involving a betrayal of trust (Koehler and Gershoff 2003). As the public generally expect ADM systems, for the sake of their algorithmic and computing advantages, to perform at a higher level of efficacy, and while they may have some stomach for trial and error, they may also be particularly frustrated by a revelation that a purportedly automated system actually involves significant human interference.

Conclusion

This chapter aimed to contribute to the scholarly discourse on ADM failures by shedding light on their contextual and cultural characteristics with the help of a simple heuristic typology. To develop effective and pragmatic policy responses, it is crucial that a contextualized and culturally-informed understanding of ADM failures be sought. This understanding will enable the formulation of policies that will often have tools already available but ones that are perhaps not yet recognized as adaptable to purportedly novel situations.

For the analysis here I primarily used examples drawn from China's experiences of widely deploying ADM systems across social, economic, and public administrative contexts. One objective here, not surprisingly, is to caution observers against the simplistic application of Western perspectives, attitudes, and discourses on ADM to the descriptive and prescriptive thinking about ADM failures in China. An obvious extension of this is recognition of the need for similar caution when critically examining endeavors to apply ADM in other non-Western societies with developmental aspirations. Moreover, even in Western societies, the ADM discourse may benefit from a greater diversity in tone than is apparent in the currently dominant apocalyptic narrative. Extant comparative studies suggest that, beyond the academic and policy elites' narratives, the general public in the West also exhibits diverse preferences towards novel data technologies associated with ADM (Kostka, Steinacker and Meckel 2021). Acknowledging, as opposed to suppressing, the more localized perspectives on ADM and their failures, and tailoring policy responses accordingly, will enable societies to better harness the decisional advantages of ADM while living with its must-be-tolerated shortcomings.

Acknowledgement

This research is supported by PKU-Wuhan Institute for Artificial Intelligence.

References

Ajunwa, Ifeoma. 2020. "The Paradox of Automation as Anti-Bias Intervention." *Cardozo Law Review* 41 (5): 1671–1742.

Amstrong, Mark. 2013. "Recent Developments in the Economics of Price Discrimination." In *Advances in Economics and Econometrics: Theory and Applications, Ninth World Congress*, by Richard Blundell, 114–126. Cambridge: Cambridge University Press.

Calo, Ryan, and Danielle Keats Citron. 2021. "The Automated Administrative State: A Crisis of Legitimacy." *Emory Law Journal* 70 (4): 797–845.

Chen, Dan. 2017. "Local Distrust and Regime Support: Sources and Effects of Political Trust in China." *Political Research Quarterly* 70 (2): 314–326.

Chesterman, Simon. 2021. *We, the Robots? Regulating Artificial Intelligence and the Limits of the Law.* Cambridge: Cambridge University Press.

Crawford, Kate, and Jason Schultz. 2014. "Big Data and Due Process: Toward a Framework to Redress Predictive Privacy Harms." *Boston College Law Review* 55 (1): 93–128.

Eubanks, Virginia. 2018. *Automating Inequality: How High-Tech Tools Profile, Police, and Punish the Poor.* NY: St. Martin's Press.

Kahneman, Daniel, Oliver Sibony, and Cass Sunstein. 2021. *Noise: A Flaw in Human Judgment.* NY: Little, Brown Spark.

Kaminski, Margot E., and Jennifer M. Urban. 2021. "The Right to Contest AI." *Columbia Law Review* 121, (7): 1957–2048.

Kleinberg, Jon, Himabindu Lakkaraju, Jure leskovec, Jens Ludwig, and Sendhil Mullaninathan. 2018. “Human Decisions and Machine Predictions.” *Quarterly Journal of Economics* 133 (1): 237–293.

Koehler, Jonathan J., and Andrew D. Gershoff. 2003. “Betrayal Aversion: When Agents of Protection Become Agents of Harm.” *Organizational Behavior and Human Decision Processes* 90 (2): 244–261.

Kostka, Genia, Lea Steinacker, and Miriam Meckel. 2021. “Between Security and Convenience: Facial Recognition Technology in the Eyes of Citizens in China, Germany, the United Kingdom, and the United States.” *Public Understanding of Science* 30 (6): 671–690.

KPMG & University of Queensland. 2023. “Trust in Artificial Intelligence: A Global Study”.

Kroll, Joshua A., Solon Barocas, Edward W. Felten, Joel R. Reidenberg, David G. Robinson, and Harlan Yu. 2017. “Accountable Algorithms.” *University of Pennsylvania Law Review* 165 (3): 633–706.

Lu, Haibo et al. 卢海波等. 2022. “AI Fu Neng + Hegui Daoxiang + Xitong Bihuan: Mangguo TV Neirong Anquan Baozhang Jishu Tixi de Goujian”AI赋能+合规导向+系统闭环:芒果TV内容安全保障技术体系的构建[AI empowered + Compliance oriented + Systematically close-looped: Constructing the Content Safety Technological System for Mango TV]. *Guangbo Dianshi Xinxi* 广播电视信息6:21–24.

McKendrick, Joe, and Andy Thurai. 2022. “AI Isn’t Ready To Make Unsupervised Decisions.” *Harvard Business Review.* September 15. https://hbr.org/2022/09/ai-isnt-ready-to-make-unsupervised-decisions.

Morozov, Evgeny. 2011. *The Net Delusion: The Dark Side of Internet Freedom.* NY: Public Affairs.

Pence, Mike. 2018. “Vice President Mike Pence’s Remarks on the Administration’s Policy Towards China.” *Hudson Institute.* October 4. Hudson Institute.

Salib, Peter N. 2022. “Big Data Affirmative Action.” *Northwestern University Law Review* 117 (3): 821–892.

Sunstein, Cass R. 2022. “Governing by Algorithm? No Noise and (Potentially) Less Bias.” *Duke Law Journal* 71 (6): 1175–1205.

Tirole, Jean. 2017. *Economics for the Common Good.* NJ: Princeton University Press.

Yan, Tian阎天. 2021. “Nvxing Jiuye zhong de Suanfa Qishi: Yuanqi, Tiaozhan yu Yingdui” 女性就业中的算法歧视:缘起、挑战与应对[Algorithmic employment discrimination against women: Origins, challenges and responses]. *Funv Yanjiu Luncong*妇女研究论丛5: 64–72.

Yu, Haiqing. 2022. “Living in the Era of Codes: A Reflection on China’s Health Code System.” *BioSocieties.*

Yu, Haiqing, and Jesper Zeuthen. 2023. “Local Politics in the Age of Automated Decision-Making in China: A Case Study of the Henan Health Code Scandal.” *Journal of Contemporary China.*

Zeng, Xun. 2007. “Enforcing Equal Employment Opportunities in China.” *University of Pennsylvania Journal of Labor and Employment Law* 9 (4): 991–1026.

Zheng, Yueping et al. 郑跃平等. 2020. “Difang Zhengfu Shuju Zhili de Xianzhuang yu Wenti—Jiyu 43 ge Zhengwu Rexian Bumen de Shizheng Yanjiu”地方政府数据治理的现状与问题——基于43个政务热线部门的实证研究[Status and problems of data governance by local governments: An empirical study of 43 government hotlines]. *Dianzi Zhengwu*电子政务 7: 66–79.

Zuboff, Shoshana. 2019. *The Age of Surveillance Capitalism: The Fight for a Human Future at the New Frontier of Power.* NY: Public Affairs.

CHAPTER 3

A Democratic Ethos? Explorations of Blockchains and Governance in China

Warwick Powell

Abstract

The chapter discusses blockchain technologies in China's economic and social governance. It suggests that blockchains with Chinese characteristics are emergent foundations of a potentially new ethos of socialist governance, which hinges on the expansion of the public sphere of data via the creation of data ecologies anchored by blockchains. Chinese ambitions of governance reform, namely those of co-production, co-governance and co-sharing, are amplified through the capability of blockchains to enlist a plurality of agents to a common informational integrity purpose, which is necessary to sustain the deliberations and reflections that take place through the consultations between the governed and the governing.

Keywords: Blockchain; Governance; Democracy; Data ecology; Common knowledge

Introduction

When it comes to questions of technology, democracy, and China, the common refrain describes China variously as undemocratic, authoritarian, or autocratic, labelling technology in China as a tool of mass surveillance in the hands of a despotic state apparatus. Similarly, the digitalization of life-worlds—encompassing how human beings frame, experience and interact with the world in which they are situated—and the role of data in China are often portrayed in dystopian terms. The spectre of state surveillance is cast as the leitmotif of contemporary digitized totalitarianism, despite the power configuration of surveillance being increasingly refracted through the willing and often enthusiastic embrace of self-surveillance by citizen subjects themselves (Bauman and Lyon 2013; Powell 2022). In the absence of mechanisms for mass plebiscites, China's digitalization footprint is presented as the sine qua non of anti-democracy, or as authoritarianism writ large. Various tropes about Chinese governmental practices and institutions presuppose that its institutions and governance arrangements—the very 'nature' of the State—are authoritarian, and by dint of that, are not democratic.

Invocations of ideas like authoritarianism or autocracy to describe China in the late 2010s and early 2020s are increasingly common. Some have argued that China's growing global assertiveness is aimed at creating a world that is safer for autocracies (Weiss 2019). Others simply label China as authoritarian, and then proceed to wonder what makes the Communist Party of China (CPC) resilient in the face of seemingly questionable legitimacy (A. J. Nathan 2017; Ang 2022). Without periodic mass plebiscites, supposedly the mechanism of legitimacy par excellence, the persistence of the Chinese state becomes a puzzle for some (A. J. Nathan 2003).

Yet, as Primo Levi (1989, 138) has warned:

> To give a name to a thing is as gratifying as giving a name to an island, but it is also dangerous: the danger consists in one's becoming convinced that all is taken care of and that once named, the phenomenon has also been explained.

And so it is when it comes to questions of technology, democracy, and China. Simply naming something does not necessarily make it so. If Levi's warning catalyzes a more critical reflective posture towards this Manichean nomenclature, the grist of this chapter's contemplation is anchored by the technology now known as 'blockchain'. Through an exploration of some features of the emergent technological landscape and modes of articulating blockchain technologies, in the context of wider considerations of governance in China, an altogether different picture reveals itself.

Blockchain technology features in discussions about the future trajectory of governance (zhili 治理) reform in China, among other things. The focus has been on enabling more effective multipolarity in the governance arrangements or configurations of information systems themselves, addressing the stated ambitions of governance reform, namely those of co-production, co-governance and co-sharing (*gongjian* 共建, *gongzhi* 共治 and *gongxiang* 共享). These three concepts are cornerstones of governance modernization and reform, articulated in the 4th Plenum of the 19th National Congress of the CPC (October 28–31, 2019). *Gongjian* focuses on collective participation in social construction and is seen as the foundation of social governance. *Gongzhi* involves collective participation in social governance, requiring a broad social perspective, aimed at opening up a governance system that enables the participation of all people. *Gongxiang* goes to the collective sharing of the results of social governance, requiring a fairer distribution of the results and benefits of social governance (CPC News, December 3, 2019).

This chapter suggests that the introduction of blockchain technologies into the configuration of governance itself, intentionally or otherwise, could lead to the expansion of the public life aspects of what I call data ecologies, with democratic possibilities. Drawing on Ranciere and Durkheim, I suggest that bringing to public life the conditions of information production, collection, validation, storage, dissemination, access, and utilization—the six elements of data ecologies—holds important democratic possibilities.

The common and mutual knowledge possibilities of socially credible information created and stored via the distributed mechanisms of blockchain-enabled information systems can contribute to addressing the corrosive elements of what Byung-Chul Han calls 'infocracy', defined as the explosion of information in contemporary social settings that can undermine the very possibilities of common truths that make stable and democratic life possible (see Powell et al. 2021). Without addressing these foundational issues of common or mutual knowledge, of shared truths, or what Baudrillard (1994) called the 'sacramental order', a focus on procedures alone no more guarantees 'democratic' outcomes than does casting a ballot at a periodic plebiscite. The introduction of blockchain technologies to governance reform in China can therefore introduce a credible basis for enhanced deliberation and reflection upon a series of common and mutual knowledges prevalent across the social body at large—or, generally speaking, enable an effective dialogue between the governed and the governing, which arguably is the sine qua non of functional democracies in a Durkheimian sense.

My core proposition, therefore, is that public ecologies of data are presaging an ethics of democratic practice in China in which the challenges of achieving and sustaining social consensus 'from the bottom up' can be addressed while at the same time squaring the circle of social stability (*zhixu* 秩序) and security (*anquan* 安全). Data ecologies are both sites of knowledge power tout court and the means by which knowledges (or information/data) can be shared over diverse sites across the socioeconomic body at large. For Foucault (2008), power is based on knowledge, reproducing knowledge by shaping it to its requirements. Knowledge is possible and takes place within networks of power relations. These networks enable knowledge to come into existence, to validate various knowledge truth claims, and allow knowledge itself to be mobilized in the name of governing conduct. Power relations determine the conditions of existence of knowledge production, validation, and use. Distributed ledger technologies, and blockchains in particular, are the means by which new power relations in the context of knowledge are being considered and designed into modalities of social and economic governance in China.

This chapter is organized into sections. Section 1 briefly introduces blockchain technology and describes its design and operational sine qua non. It discusses the main elements of the concept of data ecologies. Section 2 explores the emergence of a Chinese sensibility in relation to blockchain technologies during the 2010s with a focus on supply chains, efficiency of payments, and the importance of dependable data. Data qua public resource raises questions of democratic practice and ethos, which are the subject of Section 3. Here, the emerging discussions in

China about the possibilities of blockchains in contributing to improved social governance are explored. This section suggests that blockchains, in the context of governance reform, open up channels and vectors to reinvent democratization in senses inspired by Ranciere (2014) and Durkheim (2018). A concluding discussion explores some of the implications of the application of blockchain technologies to governance and to questions of democracy in China (and elsewhere).

Blockchain as an Element of Data Ecologies

Blockchains are information systems involving a multitude of parties, acting together, in processes through which certain information is formed, validated, stored, and transmitted. They are decentralized computer networks that are fault-tolerant and resistant to attack and collusion. As such, blockchain information systems ensure that a data state is valid because it came into being via a process defined by a set of rules that are transparent to participants; such data states are generally speaking not reversible without significant cost and difficulty (Mehar et al. 2019; Zichichi et al. 2019), while the information on the chain is censorship- and fraud-resistant, being difficult to alter singularly and capriciously. Information symmetry and synchrony are enhanced, mitigating the costs of asymmetry (Akerlof 1970). Risks of corruption are also reduced (DiRienzo et al. 2007; Barth et al. 2009).

Blockchains undergird a social information ecology in which multiple agents can be involved in the various moments of the data life cycle. This is possible because the distributed architecture of blockchains enables the mobilization of different agencies, each with discrete identities secured by way of public-private key security, and whose actions in relation to each moment can be governed by way of protocols. Each action is recorded on a ledger that is commonly operated and maintained by a network of stakeholders (that is, those with a vested interest in the integrity of the data overall) and can be audited by various stakeholders as a bulwark against informational malfeasance.

Blockchain is a 'social technology' in that it enables the involvement of a multitude of actors in various aspects of data ecologies; it also implies the existence of a social division of labor in the supply chain of data itself (Powell 2022). It is an ecology of data and relations associated with the conditions of existence of data itself. An ecology deals with the relation between things and with their broader surroundings. A data ecology therefore is a way of dealing with the relations between all those things that make up a social information system and their relations to one another. Blockchains, as distributed ledgers involving multiple actors, are a foundational element of a data ecology. The idea of data ecology brings to the fore the entire life cycle of information and its conditions of existence. Data is in effect treated within a supply chain frame, where each discrete moment of activity is subject to participatory agency, governmental consideration, and institutionalization.

Blockchain and Data in China

The adoption of blockchain technologies in China has been something of a paradox. Regulators have been uneasy about cryptocurrencies; in effect, they have been banned; yet blockchains have now been elevated to the status of one of a handful of cornerstone technologies to be developed and promoted as part of China's current phase of data-enabled economic and social development (see Powell 2022, for a detailed examination of the emergence of blockchains in China). In October 2019, President Xi Jinping announced that blockchain must be treated as an 'important breakthrough for independent innovation of core technologies' and that China must 'accelerate the development of blockchain and industrial innovation' (Xinhua, October 25, 2019). Come April 2020, the State Council formally designated data as a factor of production (*shengchan yaosu* 生产要素) alongside land, labor, capital, and technology. From 2015 onwards, academic and governmental research was being undertaken into the opportunities and relevance of blockchain to China's economic and social development needs (Kuang and Peng 2020; Qian, n.d.). In February 2016, the State Council issued the 13th Five Year National Information Plan that opened the door to blockchain technologies. This was followed by the publication of the China Blockchain Technology and Use Case Development White Paper by the Ministry of Industry and Information Technology (October 2016). The White Paper highlighted the distributed storage characteristics of blockchains, and the concomitant barriers to capricious data alteration that were thus implied. The idea of immutability (*buke cuan'gai* 不可篡改) shone like a beacon in a sociopolitical landscape in which fraud, counterfeiting, and corruption remained rife. Tackling fapiao (invoice) fraud with distributed ledgers was one of the early-use applications, with trials implemented as early as 2018 (see Powell 2022, 148–49).

In an article published on the website of The Supreme People's Procuratorate of the People's Republic of China, Tian and Zhang (2020), respectively the Deputy Director of the Integrity Research Center of Hunan University and a graduate student at the School of Public Administration of Nanchang University, identified blockchain's capacity to strengthen measures to prevent corruption, specifically referencing the capacity of blockchain-tracked data to be compared over time, thereby acting as a bulwark to capricious data malfeasance that takes place in traditional siloed information systems due to the 'low cost of changing information'. Recently, prisons in Jiangsu Province implemented a blockchain-based system to manage parole, commutation, and prisoner assessments with one of the main aims being to remove corruption in the process (see Feng 2021). According to reports, the 'whole process of law enforcement by police will leave online records and 100 per cent of the cases can be traceable [to every step]' (Jiangsu People's Government, December 11, 2021). Information integrity was, throughout, the principal focus of concern for public policy analysts and researchers as they grappled with the possibilities of blockchains for China's socioeconomic development and governance.

The data dependability and transaction efficiency properties of blockchains made the technology a glove-fit for supply chains (Sigley and Powell 2022). If the 1980s through to the late 2000s were the years of unleashing the 'forces of production', by the mid-2010s the focus had shifted to the question of capital accumulation—namely the conditions of circulation and metamorphoses of capital itself. Whereas Deng invoked the dictums of 'productive forces' that feature prominently in Capital Volume 1, the issues of circulation that comprised the core content of Capital Volume 2 started to find their place in the policies and thinking from the late 2010s (Powell 2022). Blockchain applications in supply chain traceability have been extensive, and have aimed to address issues such as food safety and the integrity of provenance claims to mitigate risks of food fraud, as documented in, for instance, Xiaowei Wang's (2020) Blockchain Chicken Farm. Opportunities for blockchains to support the development of the Belt and Road (BRI) initiative were also identified, such as in supply chain integrity, cross-border data flows, and payment efficiency fields (Liang Haiming, December 22, 2015; Lin Nuoming, December 14, 2017; Hu and Fei 2018; Lin Yan 2019; Li Yutong and Zhang Fangfang n.d.).

Blockchains were positioned as an alternative to SWIFT, the dominant inter-bank messaging service that enabled cross-border payments to be made; blockchains would reduce the costs and delays associated with SWIFT (Zhang Lei 2018). As Liu Xiaolei (2019) argued, blockchain technology deployed in supply chain transactional contexts can resolve counter-party risks while enabling the 'passing of funds with one hand, and the transfer of goods with the other' (*yishou jiaoqian, yishou jiaohuo* 一手交钱, 一手交货). Near real-time funds transfers enabled by data synchrony and symmetry would reduce the volume of capital-in-use (*zijin zhanyong* 资金占用), increase the retention of capital-at-hand (*chiyou zijin* 持有资金), reduce intermediation costs, and lower transaction fees (cf. Zhu Caihua 2020, 19; Powell et al. 2022). Improved data flows also enabled reduced delays at customs, a point made by customs officials (cf. Zhang Morong 2020), as data transparency improves supervisory capability (Zhang and Jin 2020). By mid 2023, the 'Silk Road Cloud Chain' enabled the issuance of the first blockchain electronic bill of lading pilot for China Merchant Shipping and its partners (Zhongguo Hangwu Zhoukan, July 17, 2023), while the use of blockchain in the emerging asset-backed securities market in China is already delivering improvement by way of reduced yield spreads (Liu, Shim and Zheng 2023).

Blockchains in Governance

Blockchains with Chinese Characteristics is not just a high-technology program, however. The possibilities of distributed ledgers a la blockchains have captured the imagination of Chinese scholars and policy researchers and practitioners with a wider remit—possibilities such as improved governance and contribution to the broader aim of informatization of government services as articulated in the *14th Five Year National Informatization Plan* (Central Commission for Cybersecurity and Informatization, December 28, 2021). Blockchain is referenced at least 21 times in this Plan. Aside from the economic and supply chain applications, there have been applications in areas such as the development of the tax credit system (Chen Qian 2020), community epidemic prevention (Chen Peiqi 2020), poverty alleviation (Feng and Zhang 2020), municipal governance in ethnic regions (Kang 2020), records management (Liu et al. 2020), social work and urban community governance (Tang and Wang 2020; Wang et al. 2020; Wang 2020), and community corrections (Yuan 2020).

The emergent terrain of ecologies of data in China configures data as a public resource, with responsibilities across ecologies distributed to a diverse range of actors. Here, I will focus on how blockchain technology is being conceptualized through the lens of social governance (*shehui zhili* 社会治理). In doing so, I suggest two things. First, blockchains are being mobilized in the name of improved governance, focused on enabling more effective multi-party involvement in the governance architectures themselves. This finds resonance and political imprimatur in the 2021 National Informatization Plan which articulates the need to 'promote the innovative application of blockchain, secure multi-party computing, federated learning, and other technological models in data circulation' as a way to 'deepen public data resource exploitation and use' (Central Commission for Cybersecurity and Informatization, December 28, 2021, 16). Multi-party computing and distributed ledgers are part and parcel of the expansion of the public life aspects of data ecologies.

Second, these democratic possibilities, coupled with the foundations of a common knowledge, make possible a credible basis for enhanced dialogue on an emergent ethics of democratic practice. Ranciere's radical critique of mainstream democratic discourse argues that democracy is a process, a practice, of 'challenging … governments' claims to embody the sole principle of public life and in so doing be able to circumscribe the understanding and extension of public life' (Ranciere 2014). For him, there is no limit specific to democracy understood in this sense, as democratization is the continual displacement of the limits of that which constitutes public and private. Democratic activity aims to broaden the definition of the public sphere, to move towards 'the power of anyone and everyone' (72). Institutionalized polities are more or less democratic depending on the extent to which la chose publique are monopolized by an oligarchy of political functionaries and economic powers. So-called electoral democracies are not democratic; they are 'States of oligarchic laws' in which elections ensure a revolving door of 'dominant personnel ... under interchangeable labels' (73).

Traditional information systems, the archives of state apparatuses so to speak, embody oligarchic relations of power. State apparatuses or large corporations collect and hold information, and—in the name of it being the sole principle of public life—arbitrate and regulate the authorities that make bodies of information acceptable and usable. The power of sovereignty is not restricted or reduced to power over life and death; it is the right to create and make permanent a mark on a record and to alter or erase such marks capriciously. Emergent data ecologies are, by design or otherwise, giving rise to spaces of democratic action, that is, spaces in which authority over the conditions of existence of information is rendered a public concern, as discussed by Tian and Zhang (2020) and others in relation to the Jiangsu prison management case. If information is the grist of public life, then emerging ecologies of data are framing opportunities for institutional configurations and practices that delimit the oligarchs' reach, ability to control data, or lay claim to data's valorization.

By the late 2010s and early 2020s a burgeoning discourse about blockchains and enhanced governance in China emerged (cf. Yang and Yu 2019; Yu Yuxin and Zhang Yugui 2020; Zhang Jiaxing et al. 2020; Qian and Gao 2020). This came on the back of declarations made at the 4th Plenum of the 19th National Congress that specifically highlighted the importance of improvements to socialist governance as part of China's ongoing development towards its 2035 and 2049 objectives. As Yang and Yu (2019, 47) have observed:

> 'Blockchain technology is the foundational technical layer, and also changes how the masses can participate in social governance, and improves the promotion of social coordination and democratic consultation. The distributed nature and P2P capabilities of blockchain enable multilateral agency in social governance, supporting the ability of each social agency to establish trust mechanisms under the direction of Government. Blockchain's most valuable and significant elements are found in the adjustment to the relationships between people, enabling the transformation of the historic models of social governance that saw government responsible for all aspects, to an environment in which people can participate in social governance and deepen their involvement in socio-economic development and change.'

These observations are echoed in a range of papers released during 2019 and 2020. We can immediately see from the quote above that the technical dimensions of distributed networks (viz., the number of computer nodes involved) are readily transposed into a social context of agents. Blockchain nodes are not mere technological cogs; rather, they are functional extensions of social agencies. The enlistment of social agencies into mechanisms associated with governance of informational dependability represents a radical departure from historic systems 'that saw government responsible for all aspects'. Nanjing-based think-tank scholars, Qian and Gao (2019, 22), put it this way: 'Actually, multilateral governance inherently includes the internal self-governance of multi-agent groups'. The 'self-governance of multi-agent groups' speaks to associations that form around a body of data and its sociotechnical conditions of existence—data ecologies—and the need and responsibility of members to contribute to the integrity of the data ledger upon which they all depend. Qian and Gao (2019, 26) continue:

> 'This aside, blockchain is a distributed network ledger technology, simultaneously controlled by multiple nodes, in which the ledger and its integrity are maintained and populated through participant competition. This kind of technology is at root the embodiment of a kind of openness and sharing democratic governance, in which each node expresses its authority and in accordance with the views of the majority form a unanimous outcome ...'

Blockchains are finding a place in numerous application environments because they can expand the participation of actors in the governance of the information system itself, with scalable possibilities. As Gao and Qian (2019, 30) claim:

> 'Social governance advocates for the proactive participation of multiple agents. The participation of people in social governance is an element of the democratization of national governance...'.

Blockchains are being linked to programs of 'democratization of national governance' (Yang Jun 2019; Yang Xu 2020; Yu and Zhang 2020). We can hear an echo of Ranciere here. In tangible terms, blockchains are being mobilized because they enable the dilution of previously unilateral authority in information systems: 'Decentralized is the distinguishing property of blockchain technology. On the basis of a reciprocal network, and through a consensus algorithm, *unilateral control is avoided*' (Luo Xin and Cai Yuting 2019, 24 – my emphasis). Community level multilateralism in party-building and urban governance is also seen as something that can be achieved through the decentralized properties of blockchains (Wang Deqi et al. 2020). Blockchains are therefore seen as a foundational layer in a broader techno-social project of governmental modernization in China (see for instance Cao and Hou 2020; Chen Peng 2020; Cui Yanbing 2020; Huang Li 2020; Li and Zhang 2020; Su Yu 2020; Yang and Yu 2019; Yu and Zhang 2020). This governmental modernization program seeks to align aspirations of effective administration with the 'original ambitions' of democratic socialism driven by a revivified CPC. Modernized government, with the aid of digitalization and distributed ledger technologies, aims to tackle siloed and atomized data that act as barriers to efficient, accurate, and timely administrative actions and decisions (see Feng Chaorui and Zhang Yeqing 2020 for a case study on overcoming data sharing and access problems in the roll-out of national poverty alleviation initiatives). Data silos also reinforce the presence of data territories and associated oligarchs.

The intellectual ambition is to articulate digital technology systems into power networks that give effect to the three reform ambitions: co-create, co-govern, and co-share. While levelling out the information playing field by enlisting a network of interested stakeholders as active agents in distributed data networks with responsibility for system integrity and multipolar consensus, technology is being leveraged to transform the social relations of power vis-a-vis data. System transparency is expected to mitigate the risks of malfeasance and unlawful behavior and become self-reinforcing through improved agent behaviour (Luo Xin and Cai Yuting 2019). From a data ecology perspective, blockchains are foundational infrastructure that enable the participation of a multiplicity of stakeholders in their establishment (*co-create*) and in their operations through the administration and implementation of agreed protocols (*co-govern*). The results of these protocols and data infrastructure, namely dependable and credible data, can therefore be harnessed by a diversity of stakeholders, while the benefits of data in the fields of productivity and governance can be accessed by a wider audience (*co-share*) via market and other distributive mechanisms. Behavioral governance is also transposed to the agents themselves, through self-regulation of data-related conduct in conditions of transparency and consensus-driven social accountability. Agents are both the watched and the watchers, watching themselves and watching others.

The design, operationalization, and governance of data ecologies are not, however, ends in themselves; rather, the extended benefits of credible and dependable common knowledge—social truths, so to speak—lays a foundation for functional democratic governance. Democracy in practice must be more than periodically collecting a ballot if it is to be substantially meaningful. The associationalist critiques of the late 19th and early 20th centuries make this clear (see Hirst 1988; Hirst 2005; Hirst 2013), as does the work on voting mechanisms from within Western blockchain/cryptocurrency communities.

For Durkheim (2018), a government is distinct from the individuals that form the multitudes precisely because it can adopt a different posture on the issues of the day, separate from the spontaneous and uninformed perspectives of individuals. The organs of democratic governments are necessarily more than a mere reflection of an unconditional, unreflective majority disposition; rather, these organs have the capacity to draw on the perspectives of the multitudes, mobilize new bodies of information and knowledge, and arrive at policy postures that are at once grounded in the multitude's disparate dispositions and the dissociated insights of detached rationalities. 'Deliberation and reflection ... are features of all that goes on in the organs of government', he said (79). Democratic government is not a static state, defined by the presence of some institution or another; it refers instead to a dynamic process:

> From the moment that the people set themselves the same question as the State, the State, in solving them, can no longer disregard what the people are thinking. It must be taken into account. Hence the need for a measure of consultation, regular or periodic. (81)

For Durkheim, the issue is not so much about either the number of people that have a part in the government of a community, or the mechanisms by which those in government are chosen. Electoral plebiscites are not a sign of functional or effective democracy; rather, a democratic state is one characterized by ongoing interaction between those in government and the wider community. He observes that '[t]he more that deliberation and reflection and a critical spirit play a considerable part in the course of public affairs, the more democratic a nation' (89). There are echoes here of Habermas who observed that dialogue is credible when it is based on a shared information base, a background knowledge, so to speak, that is credible and accepted by all involved (Habermas 2015). There are political and arguably cultural limits to his rationalist presuppositions, however, which I will touch on in the conclusion.

A common basis of social truths is a necessary condition of existence of a democratic polity. Without common truths, there can be no functional dialogue and deliberation, and without these there can be no functional democracy. Blockchains qua foundational information integrity infrastructure, involving the exercise of various responsibilities of different actors vis-a-vis the functions that make up data ecologies, open up the conditions of social information itself to public spheres. At the same time, the participation of a disparate network of stakeholders in the creation and sustaining of applicable 'sacramental orders' is necessary to enable the dialogue needed for functional democratic relations between those governing and those being governed.

Discussion: An Ethics of Democratic Practice?

'If we want to comprehend what kind of society we are living in, we need to understand the nature of information', says Byung-Chul Han (2022b). This chapter has explored some of the dimensions of information and its conditions of existence in the context of an emerging field of governmentality in China, that is, the data ecologies made possible by the emergence of blockchain technologies and the articulation of blockchain technologies into aspects of Chinese socioeconomic life. The digitalization of the socioeconomic milieu, as in supply chains and social governance, are therefore part and parcel of an evolution of governmental technologies that seek less to intensify the modalities of external policing than to amplify the capacities of devolved self-governance.

As China emerged from an extended period in which questions of class struggle dominated governance, policing, and surveillance to one in which 'the contract' is the organizational leitmotif of economic reform and opening-up, new values and dispositions began to be shaped. Dutton examines the organizational principles and evolution of socialist policing (Dutton 2012). Schmitt's friend-enemy dyad shapes Dutton's historical exegesis in which he argues that socialist policing was underpinned—at least until reform and opening-up (from 1978 onwards)—by the need to distinguish friend from enemy. Schmitt's 1927 political dyad paralleled Mao's 1926 dictum: the pertinent question, Mao asked, is 'who are our friends and who are our enemies?'. Class struggle was thus the fulcrum for policing and for the classification of subjectivities according to this overarching political logic. Class struggle and the mass line, in which managerial or professional expertise was eschewed, were also the foundations of new management systems in enterprises during the Cultural Revolution, as ideological and political calculus was imposed, overtaking the previous emphasis on technocratic divisions of labor and the use of financial bonuses to drive performance (Bettelheim 1974). Come 1978, however, and the logic and passions of class struggle began to give way to new imperatives.

Dutton documents in detail the impact of commodification of information on policing in the post-reform and opening-up period, including the payment of 'snitches' for information from the streets. He argues that commodification of life has made the world calculable: '... the contract makes things calculable or economically viable or visible', but also 'produces new mentalities and value systems that lead to the reification of money and the commodity economy' (1992, 295). Our story picked up on this observation and sought to plug a gap or two. If, for Dutton, it is the contract that 'makes things calculable or ... visible', I have suggested that in fact the contract only frames what needs to be calculable, while it is information exogenous to the contract that makes the contract a meaningful performative instrument. Contracts may stipulate performative conditions, against which behaviors and results are judged and rewards or punishments meted out, but how is it that behaviors and results are enumerated and evaluated? This can only be done against a body of information that is considered acceptable, credible, and appropriate to the conditions of the contract itself and to the parties involved. Without information, there is actually no way that a contract's ambitions to render the world 'calculable' or 'visible' can be achieved.

The body of information is epitomized in the role that blockchains play in data ecologies. As discussed earlier, blockchains are a social technology, enabling the enlistment of a multiplicity of stakeholders to play roles throughout the entire supply chain of data. A data ecology lends itself to a social governance agenda framed by gongjian, gongzhi and gongxiang. The emerging Chinese scholarship on the role of blockchain technologies in the digitalization of economic and social life, coupled with a range of institutionalized deliberative practices and mechanisms, begins to define what I suggest could be called a 'democratic socialist governmentality with Chinese characteristics'.

There are some caveats to such an intellectual construction. First, the effects of the design, application, and institutional integration of blockchains are contingent on the wider context. As a foundational technology, blockchains in themselves create the conditions of possibility for a wide array of operational and political effects. In supply chains, the operational intent and likely effect is the streamlining of capital metamorphoses, rendering economic activity and agents mere aliquots of a greater, increasingly autonomous machine. This in itself provokes questions about the future role of human creativity and labor itself. As a foundational information system that makes alterations to key design parameters difficult, blockchains also hold the potential to lock in design deficiencies that are costly to address. How these 'on-chain' protocols are integrated with off-chain governance regimes remains pivotal to the capacity of blockchains to enhance the conditions of existence of data ecologies and, perhaps, create the conditions necessary for democratic proximity between the governed and the governing. Dimensions of polarity and the distribution of authorities in data ecologies remain open, political questions.

Second, the democratic trope that dominates Western claims about its own virtuosity provokes a more considered reflection of questions concerning distributed authority. Democracy is not reducible to periodic plebiscites, while digitalized technologies are not necessarily the pernicious extensions of an authoritarian state hell-bent on coercive control, surveillance, and punishment. Data ecologies not only create the conditions in which the systems of information themselves can become democratized in Ranciere's sense, but they also provide the grist that activates 'deliberation and reflection'. It is both a condition of existence for democratic practice and a series of sites for public oversight.

Third, effective democratic practice is fluid. Ranciere reminds us that democracy is the practice of enlarging the public sphere as a bulwark against acts of governmental and private oligarchs that seek to shrink it, 'making it into its own private affair and, in so doing, relegating the inventions and sites of intervention of non-State actors to the private domain' (55). Enlarging the public sphere, however, is not tantamount to inviting State encroachment on society; rather, it is a struggle to counter the distribution of the public and private in ways that consolidate the dominance of oligarchs in both state and civil society. The enlargement of the public sphere involves extending recognition of the public character of types of spaces and relations that were once left to the discretion of the powerful and the wealthy. In this sense, the ever-enlarging spaces that are considered 'public' are consistent with the notion of functional self-governing communities, with these communities and their members framing what constitutes public spaces and resources and determining how best to administer, govern, and utilize them. Whether blockchains deliver on the possibilities of the expansion of the public sphere remains an empirical matter as much as it is a conceptual question.

Information is a critical resource, economically and socially. Information symmetry and synchrony condition the possibilities of actions and transactions, to the advantage of some over others. Credible and dependable information makes discourse and exchange possible. The distribution of authorities in relation to the entire fabric of information qua resource embodies a series of sites and relationships that can either be privatized by the State or by corporations, or be rendered public in Ranciere's radical sense. Democratization of digitalized life-worlds involves transgressing the limitations of access to the resources of data ecologies themselves, to recognize that an ecology and its constituent elements are public domains of life-in-common. The question of democracy in the governance of digitalized life-worlds can only be understood in the context of institutional configurations and operational practices that expand the public domain of data and which enable the transcendence of privatized authority. Such a transcendence presupposes a radical transparency across a growing array of ecologies of data, their instances and respective dimensions.

Conclusion

China's emerging statecraft sees information and data ecologies as foundational resources and as a condition of its existence. Information is conceived as a public good. Data ecologies open up many vectors for democratic expansion. Data exchanges bring information from the private domain into the public domain to not only support and enable exchange of commodities but to be the commodity of exchange itself. Blockchains, according to Chinese scholars in public policy and governance, are a critical enabling technology in China's social and political modernization.

Conceptually, I have drawn on some traditions in critical democratic discourse—particularly those Western threads that long ago critically evaluated the pretenses of liberal parliamentarism—to describe more richly the kinds of institutional designs, governance mechanisms, and dialogic practices that are necessary to create and sustain social arrangements that will support a *democracy-to-be*. In doing so, I have explored ways of bringing these sensibilities to bear in the case of Chinese governmentality and practice, with particular focus on the emergent space of data ecologies underpinned by blockchain technologies. Rather than accept the *a priori* of digitalized authoritarianism, I have instead sought to explore, in a preliminary way, the complex, multifaceted dimensions of digitalized life-worlds through the parsimonious conceptual apparatus of data ecologies. Blockchains in the context of social governance reform in China open a space in which multipolarity is being introduced into different data ecologies. Distributed ledger ecosystems or communities of common information stakeholders are, in this sense, information associations with distinct operating features in relation to the kinds of participants involved and the protocols that govern their interaction with the various aspects of data ecologies. How this emergent set of off-chain configurations nestles with or buttresses against existing state apparatuses will require further study, as real cases begin to be rolled out in earnest and at scale.

Blockchain-enabled data ecologies are not restricted to applications in the sphere of capital circulation alone; rather, blockchains speak to concerns about the conditions of existence of shared information as a basis for cultivating common truths and sustaining binding values, ideals, and convictions. Without these conditions, the digitalization of information runs the risk of fragmenting communities and undermining the possibilities of democratic life by causing the social fabric to collapse into fragmented contingencies. As Byung-Chul Han (April 21, 2022) argues: 'We cannot build a stable community or democracy on a mass of contingencies. Democracy requires binding values and ideals, and shared convictions. Today, democracy gives ways to infocracy'. Distributed information networks in which participants are collectively responsible for the veracity and validity of shared data have the potential to be an antidote to infocracy. Han may have an aversion to the spiritless effects of transparency (Han 2015), but without it the very foundations of informationalized societies would not be amenable to a new ethics of democratic practice.

Does the development of public data ecologies portend the 'end of politics'? Not so fast. Such a depoliticized democracy falls victim to what Schmitt called political romanticism (Schmitt 2017). For Schmitt, as it is for Hannah Arendt (Arendt 1960), politics is the sine qua non of human societies, encapsulating the moment when the push-and-shove of competing interests reaches its flashpoint. This political apogee, when the friend-enemy dyad comes to life, is an ever-present possibility that cannot be eradicated by dialogue. Here, Habermas' ahistorical rationalism runs up against its cultural and contextual limits. What can be reasoned in one context does not necessarily fly in another. Dialogue defers the decisive moment, but it can never eradicate the conditions of existence of the *RealPolitik* potentiality of friend-enemy relations. If the *political* is intrinsic to the human condition, the question is: what are the sites of 'the political' in data ecologies? Every vector in each data ecology is a site of democracy-to-be, and that means they are each a site of political consideration in which questions of friends and enemies are never far away.

References

Bettelheim, Charles. 1974. *Cultural Revolution and Industrial Organization in China: Changes in Management and the Division of Labor*. Monthly Review Press

Cao Haijun and Hou Tiantian. 2020. "Blockchain as a mobilizing force in social governance innovation: scrutinising value, possible challenge and outlook" 区块链技术驱动社会治理创新：价值审视，可能挑战与路径展. *Dongnan Xueshu* 东南学术, Volume 4

Central Commission for Cybersecurity and Informatization, December 28, 2021. *14th Five-Year Plan for National Informatization*. "十四五"国家信息化规划. https://www.gov.cn/xinwen/2021-12/28/5664873/files/1760823a103e4d75ac681564fe481af4.pdf

Chen Peng. 2020. "Governmental governance with blockchain embedded: raising capacity and watching out for risks" 区块链嵌入下的政府治理：能力提升与风险防范. *Journal of Guangdong Institute of Public Administration*, Volume 32 (5):13 –20

Chen Qian. 2020. "Legal approach to the construction of tax credit system under blockchain technology mode" 区块链技术模式下纳税信用体系建设法治化近路. Lanzhou Xuekan 兰州学刊, November 2020, 86–99.

China State Council. April 9, 2020. "China State Council Opinion on the Establishment and Further Improvement of Factor Marketisation and Allocative Mechanisms" 中共中央 国务院关于构建更加完善的要素市场化配置体制机制的意见. https://www.gov.cn/zhengce/2020-04/09/content_5500622.htm

China State Council. March 7, 2023. "China to Establish National Data Bureau". https://english.www.gov.cn/news/topnews/202303/07/content_WS640701c0c6d0a757729e7d87.html

CPC News, December 3, 2019. "FAQ: How to recognize a social governance system with co-creation, co-governance, and co-sharing?' 要点问答：如何认识共建共治共享的社会治理制度，http://theory.people.com.cn/n1/2019/1203/c40531-31486561.html

Cui Yanbing. 2020. "Theoretical considerations on the mobilization force of blockchain technology on innovation in social governance" 区块链技术驱动社会治理创新的理论考察. *Keji Feng* 科技风, 7–8.

Deng Peiqi. 2020. "Further analysis of the capacity of blockchain technology to assist and strengthen community epidemic prevention governance" 区块链技术助力社区防疫治理能力提升研究. *Dangdai Jingji*, Issue 4, 12–16.

DiRienzo, Cassandra E., Jayoti Das, Kathryn T. Cort, and John Burbridge. 2007. "Corruption and the Role of Information." *Journal of International Business Studies*. https://doi.org/10.1057/palgrave.jibs.8400262.

Dutton, Michael. 2012. *Policing Chinese Politics*. Durham, NC: Duke University Press. https://www.dukeupress.edu/policing-chinese-politics/.

Feng Chaorui and Yeqing Zhang. 2020. "Research on Function Optimization Path of National Poverty Alleviation Information System – A Multi-case Analysis Based on Blockchain Perspective" 区块链技术驱动社会治理信息系统功能优化—基于社会治理信息系统的多案例研究. *Journal of Guangxi Normal University (Philosophy and Social Sciences)* Volume 56 (6): 70–84.

Feng, Coco. December 15, 2021. "Chinese Province Applies Blockchain in 'Smart Prison' System to Cut Corruption and Abuse Cases". *South China Morning Post*. https://www.scmp.com/tech/policy/article/3159799/chinese-province-applies-blockchain-smart-prison-system-cut-corruption

Foucault, Michel. 2008. *The Birth of Biopolitics*. MacMillan.

Habermas, Jürgen. 2015. *The Theory of Communicative Action: Reason and the Rationalization of Society, Volume 1*. John Wiley & Sons.

Han, Byung-Chul. 2015. *The Transparency Society*. Stanford University Press.

———. 2022. *Infocracy: Digitization and the Crisis of Democracy*. John Wiley & Sons.

Hirst, Paul. 1988. "Associational Socialism in a Pluralist State." *JL & Soc'y* 15: 139.

———. 2013. *Associative Democracy: New Forms of Economic and Social Governance*. John Wiley & Sons.

Hirst, Paul Q. 2005. *The Pluralist Theory of the State: Selected Writings of G.d.h. Cole, J.n. Figgis and H.j. Laski*. Edited by Paul Q. Hirst. London, England: Routledge. https://doi.org/10.4324/9780203986004.

Huang Li. 2020. "Blockchain thinking about empowering the base layer of governance" 区块链思维赋能基层治理, in Jingji, Issue No. 24, 29-31.

Hu, Qing and Fei, Yu. 2018. "Jiyu qukuailian de kuaijing zhifu tixi ji anquan wenti yanjiu" 基于区块链的跨境支付体系及安全问题研究. Finance Wangshang Jinrong 网上金融 52–54.

Jiangsu People's Government. December 11, 2021. "Full implementation of law enforcement and case handling across all business processes; our province has established 21 'smart prisons'." 执法办案全业务全流程上线 我省建成21所"智慧监狱. https://www.jiangsu.gov.cn/art/2021/12/11/art_60085_10188817.html

Kang Lanping. 2020. "The institutional logic and innovative path of blockchain technology empowering the modernization of municipal social governance in ethnic regions" 区块链技术赋能民族地区市域社会治理现代化的制度逻辑与创新路径. *Guangxi Minzu Yanjiu* 广西民族研究, Issue 5, 110–118.

Kuang Jinsong and Peng Weinbin. 2020. "Blockchain technology drives the development of digital economy: theoretical logic and strategic orientation" 区块链技术驱动数字经济发展:理论逻辑与战略取向. Shehui Kexue 社会科学 Volume 9, 64 –72.

Levi, Primo. 1989. *Other People's Trades*. Translated by Raymond Rosenthal, Summit Books.

Li, Yan. 2019. "Exploring innovative research on cross-border payments in the 'Belt and Road' region based on blockchain technology" 基于区块链技术 探索"一带一路"区域跨境支付的创新研究 in *Jingji Zongheng* 经济纵横 Issue 23, 140–141.

Li Yan and Zhang Ke. 2020. "Influence on the innovation direction of social governance by 'block chain' thinking'", 区块链思维对社会治理创新方向的影响. *Journal of Heihe University* 黑河学院学报, No. 11, 65–70.

Li, Yutong and Zhang, Fangfang (n.d.) "Problems and solutions for cross-border payments based on blockchain technology" 基于区块链技术的跨境支付存在问题及解决建议. Xiandai Jingji Xinxi 现代经济信息.

Lin Nuoming. 2017. "Use blockchain as the backbone of the belt and road initiative" 用区块链做一带一路的骨干, December 14. https://finance.sina.cn/2017-12-14/detail-ifypsqiz6926723.d.html?from=wap.

Liu, Jing, Shim Ilhyock and Zheng, Yanfeng. August 2023. *Absolute Blockchain Strength? Evidence from the ABS Market in China*. BIS Working Papers No. 1116. Bureau of International Settlements.

Liu, Xiaolei. 2019. "What is the strategic significance of the digital currency that is being pushed through by the Central Bank?" 央行拟加快推出的数字货币有何战略意义?http://www.bri.pku.edu.cn/info/1065/2424.htm

Liu Yuenan, Zhang Yifeng, Wu Yupeng and Zheng Chong. 2020. "Blockchain technology and records management: A bidirectional perspective of technology and management" 区块链技术与文件档案管理:技术和管理的双向思考, 4–12.

Luo Xin and Cai Yuting. 2019. "The impact mechanism of blockchain in network society governance" 区块链在网络社会治理中的作用机制. *Journal of Guangzhou University* 广州大学学报, No. 1, 23–29.

Mehar, Muhammad Izhar, Charles Louis Shier, Alana Giambattista, Elgar Gong, Gabrielle Fletcher, Ryan Sanayhie, Henry M. Kim, and Marek Laskowski. 2019. "Understanding a Revolutionary and Flawed Grand Experiment in Blockchain: The DAO Attack." *Journal of Cases on Information Technology (JCIT)* 21 (1): 19–32.

Nancy, Jean-Luc. 2008. "The Being-with of Being-There." *Continental Philosophy Review* 41 (1): 1–15.

Nathan, Andrew J. 2003. "Authoritarian Resilience." *Journal of Democracy*. https://doi.org/10.1353/jod.2003.0019.

Nathan, Andrew J. 2017. "China's Changing of the Guard: Authoritarian Resilience." In *Critical Readings on the Communist Party of China (4 Vols. Set)*, 86–99. Brill.

Powell, W. 2022. "China, Trust and Digital Supply Chains: Dynamics of a Zero Trust World." https://books.google.ca/books?hl=en&lr=&id=Xax6EAAAQBAJ&oi=fnd&pg=PT8&ots=1lByv1WnC-&sig=ub-WuEhJyKhBYlCp6owg9KDjEBB4.

Powell, W., S. Cao, T. Miller, M. Foth, and X. Boyen. 2021. "From Premise to Practice of Social Consensus: How to Agree on Common Knowledge in Blockchain-Enabled Supply Chains." *Computer Networks*. https://www.sciencedirect.com/science/article/pii/S1389128621004606.

Qian, J. n.d. "Experimentalism as a Policy Style: The Case of China." In *The Routledge Handbook of Policy Styles*. https://doi.org/10.4324/9780429286322-6/experimentalism-policy-style-jiwei-qian.

Qian Zaijian and Gao Xiaoxia. 2019. "Blockchain, where to for 'chains': public affairs, party building and social governance" 区块链, "链" 向何方: 公共事务, 党的建设与社会治理. Governance 国家治理, Issue No. 3, 20–33.

Ranciere, Jacques. 2014. *Hatred of Democracy*. Verso Books.

Schmitt, Carl. 2017. *Political Romanticism*. Routledge.

Schuback, Marcia Sá Cavalcante, and Jean-Luc Nancy. 2013. *Being with the Without*. Axl Books.

Shi Chao. 2020. "The governance logic of trust creation and its application of blockchain technology" 区块链技术的信任制造及其应用的治理逻辑. Dongfang Faxue 东方法学, Volume 1, 108–122.

Sigley, Gary, and Warwick Powell. 2022. "Governing the Digital Economy: An Exploration of Blockchains with Chinese Characteristics." *Journal of Contemporary Asia*, August, 1–20.

Su Yu. 2020. "The Governmental Responsibility of Blockchain Governance" 区块链治理的政府责任. *Studies in Law and Business* 法商研究 Volume 37 (4), 59–72.

Tang Li and Wang Hongbo. 2020. "Application Space and Conception of Blockchain Technology in Social Work Service" 区块链技术在社会工作服务中的运用空间与构想. *Inner Mongolia Social Sciences* 内蒙古社会科学. Volume 41 (1), 146–152.

Tang Wenxian and Hu Yafen. 2020. "Application of Blockchain in Urban Governance: Value, Empowerment and Path" 区块链在城市治理中的应用: 价值, 赋能与路径. *Social Governance* 城市社会治理, No. 6, 92–102.

Tian Xiangbo and Zhang Ziwei. July 14, 2020. "The new contribution of blockchain technologies to deepen the struggle against corruption" 区块链技术为深化反腐败斗争提供新助力. https://www.spp.gov.cn/dj/llyj/202007/t20200714_472965.shtml

Wang Deqi, Lin Xiaoli and He Xiaoyan. 2020. "'Party Building + Blockchain": A New Path of Urban Community Governance" 党建+区块链: 城市社区治理新路径. *Journal of Urban Studies* 城市学刊. Vol. 41 (6): 64 –68.

Wang Xiaowei. 2020. *Blockchain Chicken Farm: And Other Stories of Tech in China's Countryside*. FSG Original x Logic.

Wang Xuezhu. 2020. "An Explanation of the Use Case for Blockchain Technology in the Context of Social Governance in the Capital" 区块链技术对首都社会治理的应用场景解析. *Shoudu Zhili* 首都治理, 75–78.

Weber, Isabella M. 2021. *How China Escaped Shock Therapy: The Market Reform Debate*. Routledge.

Weiss, Jessica Chen. 2019. "A World Safe for Autocracy: China's Rise and the Future of Global Politics." *Foreign Affairs* 98: 92.

Yang Dong and Yu Chenhui. Nov. 2019. "Blockchain Technology Applied in Governmental Governance, Social Governance and Party Building" 区块链技术在政府治理, 社会治理和党的建设中的应用. *Governance* 国家治理, Issue No. 3, 43–48.

Yang Jun. 2019. "Analysis of the Innovative Mechanism of Blockchain Technology's Assistance Towards Social Governance Modernisation" 区块链技术协助社会治理现代化的创新机理分析. *Zhongguo Hangkong Bao* 中国航空报, December 2019, Issue 7.

Yang Xu. 2020. “On the Blockchain-Driven Movement of Basic Social Contradictions and the Transformation of Social Structure” 区块链推动社会基本矛盾运动与社会结构变革. *Studies in Mao Zedong and Deng Xiaoping Theory* 毛泽东邓小平理论研究, Volume 1, 17–25

Yu Yuxin and Zhang Yugui. 2020. “Blockchain Provides Technical Support for the Modernisation of National Governance System and Governance Capacity” 区块链为国家治理体系与治理能力现代化提供技术支撑. *Shanghai Economic Journal* 上海经济研究 Volume 1, 86–94.

Yuan Jiantao. 2020. “The Application of Blockchain in Community Correction: Exploration Practice and Policy Suggestions” 区块链在社区矫正中的运用:探索实现和政策建议. *Journal of Shaoyang University* 邵阳学院学报, Volume 9 (4): 28.

Zhang Chenggang. 2018. “The Era of Blockchain: Technology Development, Social Progress and Risks and Challenges” 区块链时代:技术发展, 社会变革及风险挑战. *Frontiers*, June 2018, 33–43.

Zhang, Lei. 2018. “Research on the application of blockchain in cross-border clearing of commercial banks—the case study of China Merchants Bank’s blockchain platform” 区块链在商业银行跨境清算中的应用研究 – 以招商硬汉区块链平台为例.” PhD Dissertation, Hubei Financial University.

Zhang Jixing, Xie Yi and Peng Kaiping. 2020. “Blockchain and Social Governance: Combination, Strength and Risk” 区块链与社会治理:契合, 优势与风险. *Journal of Social Development* 社会发展研究 Volume 1, 23–37.

Zhang, Murong. 2020. “Cross-Border E-Commerce Supervision Model Innovation in The Perspective of Blockchain.” In *Reform & Innovation*, 78–80. China Academic Journal Electronic Publishing House.

Zhang, Xi, and Jin Chunyang. 2020. “Application Scenarios and Legal Risk Management of Blockchain in International Trade.” *Journal of Chang’An University* 22 (5): 20–28.

Zhao Jinxu and Meng Tianguang. 2019. “How Can Blockchains Reshape the Structure and Model of Governance?” 区块链如何重塑治理结构与模式. *Social Sciences Digest*, November 2019, 5–7.

Zhongguo Hangwu Zhoukan. July 17, 2023. “China Merchants Shipping successfully piloted its first blockchain electronic bill of lading” 招商轮船首张区块链电子提单试点成功, https://mp.weixin.qq.com/s/738kcMaqIz_CjQPUHf4OnQ

Zhu, Caihua. 2020. “Research on the legal path of blockchain cross-border payment” 区块链跨境支付法律路径研究.” PhD Dissertation, Lanzhou University.

Zhu Wanqing. 2020. “Theoretical Considerations on the Mobilisation Force of Blockchain Technology on Innovation in Social Governance” 区块链技术驱动社会治理创新的理论考察. *E-Government* 电子政务, Volume 3, 41–53.

Zichichi, Mirko, Michele Contu, Stefano Ferretti, and Gabriele D’Angelo. 2019. “LikeStarter: A Smart-Contract Based Social DAO for Crowdfunding.” In *IEEE INFOCOM 2019 – IEEE Conference on Computer Communications Workshops (INFOCOM WKSHPS)*, 313–18.

CHAPTER 4

Techno-Utopia or Techno-Trap? Unveiling the Enigma of Smart Courts in China's Judicial Reform

Fan Yang

Abstract

This chapter explores the impact of China's smart courts on diverse judicial reform measures. Through studies on four provincial data-driven systems in Chinese courts, it illustrates the limited and even hindering effect of smart courts that were originally envisioned as complements to China's judicial reform. While contributing to the professionalism of judges by ensuring efficiency, consistency, and procedural compliance while addressing misconduct, smart courts appear to focus on surface-level issues. Moreover, their implementation has led to increased centralization within the court system, deviating from delivering judicial autonomy and a "trial-centered" criminal procedure reform. They also represent an accountability dilemma for judges. Ultimately, smart courts reinforce the Party's authority and control over the judiciary and its actors through the automation of multi-layered party oversight mechanisms.

Keywords: Smart courts; Chinese law; AI; Court system; Legal tech; Judicial reform

Introduction

The People's Republic of China (the PRC) has proactively embraced cutting-edge technologies and vigorously endeavored to integrate technological advancements into its court system. While similar efforts to digitalize court systems are observed in other jurisdictions, the PRC has distinguished itself through its unwavering commitment to this transformative process. At the forefront of this undertaking is the Supreme People's Court (the SPC) which has embarked on an ambitious initiative to establish "smart courts (智慧法院)" and to effect comprehensive modernization across the entire judicial landscape through the deployment of diverse technological innovations.

Chinese courts have adopted a comparatively progressive attitude towards digitalization and automation (Liu 2019). The informatization of the judiciary in the PRC dates back to the 1990s (Deng 2021). On January 29, 2016, the term "smart court" first appeared at an SPC meeting (SPC 2016). Ever since then, the pace of digitalization has accelerated, with various data-driven systems flourishing like mushrooms after rain. The term "smart court" includes incredibly diverse and fragmented initiatives aimed at automating and digitizing the judicial process (Stern et al. 2021). The SPC describes the term as a form of organization, construction, and operation in which courts make full use of programs to support online handling of all business and to ensure openness of the whole process to judges, disputing parties, and the public (SPC 2017). One commentator defined smart courts as "legal courts where the majority of or all stages of the judicial process take place in an (online) digital environment, where some, but not necessarily all, tasks are automated with programmes that may or may not be using learning algorithms" (Papagianneas 2022). This chapter suggests, however, that the data-driven systems extend beyond the conventional scope of judicial processes to encompass all stages of court operations, notably the oversight of judicial processes by local supervision commissions (SCs) and commissions for discipline inspection (CDIs).[21] This issue will be addressed in this section.

Existing literature has been devoted to scrutinizing the conceptual framework and current state of smart court development, while engaging in robust debate concerning the implications of digitalization and automation within judicial systems. Within the distinctive context of China's political-legal landscape, smart courts are construed as instruments that bolster the Party-state's control over the judiciary and secure its centralization (Ahl and Sprick 2017 ; Stern et al. 2021). Indeed, scholarly discourse asserts that the strategic implementation of a public database for court decisions is a mechanism to enhance the regime's legitimacy by meticulously presenting a curated portrayal of the judicial process (Ahl and Sprick 2017).

Previous scholarship has shed light on the contextual underpinnings, multifaceted measures, objectives, and intricacies that characterize current judicial reform. Under Xi, judicial reforms have emphasized professionalism, legal procedures, autonomous decision-making, judges' accountability, and judiciary centralization (Ahl 2021; Li 2016); however, the actual impact of smart courts on judicial reform is not yet clear. The SPC perceives smart courts as offering valuable augmentation of ongoing judicial reform, while scholarly views on the impact of smart courts on judicial reform are diverse and inconclusive. A cohort of scholars align themselves with the perspective of the SPC, believing that technology will bring a certain degree of improvement in both the speed and effectiveness of judicial reform (Li 2021). One publication argues that smart courts enhance judicial responsibility reform, particularly reform of trial supervision and management (Papagianneas 2023). The perspectives of dissent regarding the SPC primarily focus on a singular reform agenda, that of judicial accountability. They fear that smart courts may enable judges to evade responsibility by attributing errors to AI and by claiming

[21] CDI is a party institution.

collective responsibility with technology (Gao 2022; Zhang 2022; Wang 2022; Ma 2018; Zheng 2020; Gao 2019; Papagianneas 2023).

The aforementioned studies fail to provide an in-depth analysis of emergent practices in local experiments, with their commentary remaining mostly at a conceptual level without sufficiently examining the actual impact of smart courts within the framework of post-2013 judicial reform in China, or assessing their influence on the ongoing reform process. This chapter is based on the first close examination of the impact of smart courts on various judicial reform measures. It challenges the official assertion that smart courts serve to augment judicial reform. The ultimate argument is that smart courts, despite appearing to assist in reform, actually impede and disrupt judicial reform. Smart courts further complicate and hinder an already complex judicial reform process.

This chapter presents four case studies, based on the *Annual Reports on Informatization of Chinese Courts* (No.2, 3 and 6). These annual reports offer comprehensive evaluations and summaries of diverse and cutting-edge smart court initiatives across multiple provinces and cities. Authored by research groups examining different levels of courts, these reports represent valuable sources for analysis. This chapter has further benefitted from the inclusion of sources such as news articles, official documents, and press releases, enriching the depth and scope of the research.

The remainder of the chapter is organized as follows: Section 2 provides an introductory overview of the context, methodologies, objectives, and complexity of post-2013 judicial reform, examining its correlation with smart courts in contemporary legal scholarship; Section 3 discusses four case studies, offering a precise analysis of provincial pilots of smart courts; and, drawing from these case studies, Section 4 delves into the interplay between smart courts and judicial reform. The chapter concludes with Section 5.

Post-2013 Judicial Reform

Recent judicial reform addresses professionalism, legal procedures, judges' autonomy, accountability, and centralization. The SPC and scholars favoring smart courts anticipate that they will support reform, while others fear that they may impede accountability improvements (Li 2021; Gao 2022; Zhang 2022; Wang 2022; Ma 2018; Zheng 2020; Gao 2019; Papagianneas 2023).

Juggling ambitions: the complex terrain of Chinese judicial reforms

A vast amount of academic literature has examined the context and content of recent court reforms. Before the current reforms, the decisions of adjudicators had to be assessed through a hierarchical approval process, tantamount to a system of hierarchical supervision, which reflected the perception of Chinese judges among decision-makers—that their professionalism was lacking (Li 2014). Under Xi, judicial reforms have emphasized professionalism, legal procedures, autonomous decision-making, judges' accountability, and judiciary centralization (Ahl 2021; Li 2016).

One reform element is judge quotas (员额制改革), aimed at building an elite profession of judges by admitting a limited but more capable number of people onto the judge's career path (Sun and Fu 2022). Research has identified an opposite effect, however, with elite judges being excluded by court officials, or competent judges resigning due to the workload and pressure (Song 2017; Chen and Bai 2016). It should be noted that the professionalism of Chinese judges depends not only on legal knowledge, moral standards, and judicial experience, but also on factors in the environment in which they work. To enhance judicial quality, changes in such environments, including in related professional fields, are crucial (Li 2014).

Another reform measure is judges' accountability reform (司法责任制改革) which proposes that "he who hears the case shall make the judgment and be liable therefore". While judges are allowed to make most decisions free of their supervisors' approval, they are held responsible for their decisions for life.[22] This reform has significantly enhanced judges' autonomy and accountability, reduced collective accountability and corruption, and also diminished judges' capacity to shirk responsibility (He 2021; Wang and Li 2014; Chen 2017; Wang 2020). It also marks the end of the case-approval system, reducing administrative influence within individual courts—court officials have lost the power to decide on and sign off on cases (Papagianneas 2023; He 2021). This measure fosters a distinct judicial identity separate from the political and administrative arms of government, empowering judges on the frontline and boosting their confidence in resisting interference (Sun and Fu 2022). This devolution of power does not apply in all cases, however. Whenever a case is deemed sensitive, court leaders can become involved, keeping track of progress (Papagianneas 2023). Additionally, administrative influence in individual courts remains prevalent, with court officials still able to exert influence through both formal and informal channels (He 2021; Wang 2020).

It is well known that in practice the SPC does not command unquestioned allegiance from its lower, grassroots, subordinate courts, although it is supposed

[22] The judge is liable for illegal adjudication if he/she has wilfully violated the law in the trial, or if he/she has rendered a wrong judgment causing serious consequences due to gross negligence. Chinese judges were not held accountable for their cases for life prior to this reform. Supervisors are liable for their management and supervision if they fail to supervise or conduct improper supervision intentionally or with gross negligence, resulting in wrong judgments and serious consequences. In the past, senior judicial officials were not the judges who tried the cases, so they were not held accountable.

to exercise control over them (Ng and He 2017). Moves towards centralization have therefore been pursued—called "soft centralisation"—moves that have brought court budgets and local personnel issues under provincial-level control, away from the former local level. Such centralization is aimed at countering local protectionism and avoiding costs at the national level (Ahl 2021; Mertha 2015). Optimists argue that these attempts have made the higher-level courts' control over the lower-level courts and judges more comprehensive, quantitatively based, and effective (He 2021); however, others have demonstrated that only a few provinces have achieved total provincial control of court finances, so the actual impact of this attempt is not clear (Ahl 2021; Mertha 2015; Jiang 2016). Empirical research has discovered that reforms have achieved limited results since local party-states still hold onto their influence when it comes to appointments and court budgets because of manpower, local knowledge, and financial resource issues (Wang 2021). Structural dependence of the local judiciary on the same-level party-state has even been exacerbated in some cases (Meng 2023).

The Chinese criminal process privileges "investigation-centralism", emphasizing substance over procedure (Ahl 2021; Wang and Li 2014; Guo 2021; Chen and Long 2013). The courts rarely question evidence submitted by the police or procuratorates in criminal proceedings because the institutional relationship between police departments, the procuratorates, and the courts operates as an "iron triangle", with the whole criminal process like an "assembly line". Instead of checking the work of previous parties involved in bringing a criminal charge, the Chinese police, procuratorates, and courts divide the labor involved in getting a conviction (Guo 2021; Ji and Xi 2022; Wang and Tang 2021) such that illegal evidence is difficult to exclude. Endeavors have therefore been undertaken to strengthen procedural protections and promote "trial-centeredness" criminal procedure reform. They seek to move hearings to the center of trials, guarantee that evidence is collected and used in accordance with legal standards, and that courtroom hearings play a decisive role in ascertaining facts and identifying evidence (Ahl 2021; CCPCC 2014); however, much of the literature indicates that this will require an adjustment in institutional relationships, an adjustment unlikely to occur any time soon (Guo 2021; Long 2015; Chen 2016; Wei 2020; Zhang 2015). The reform has consequently seen limited success.[23]

Research agrees that judicial reform in the PRC is bound to be arduous, subject always to processes of negotiation, compromise, and adjustment (Li 2014; Sun and Fu 2022; Ji et al. 2016). Alongside the hurdles mentioned above, judicial reform suffers

[23] There are other reform routes that scholarship investigated. For the purpose of this study, we have excluded other reform paths. These exclusions enable a more targeted and rigorous investigation into the specific research question of this study.

from several other hindrances, including restrictions determined by the political system (Ji et al. 2016; Lubman 2000; Liebman 2007; Chan 2015), constraints arising from market and social forces (Ji et al. 2016; Cheng and Li 2012), the "fragmentation" of power structures (Ji et al. 2016; Lubman 2000; Lieberthal and Lampton 1992), and the failure to consult multiple stakeholders (Ji et al. 2016; Sun and Fu 2022). Moreover, the establishment of a National Supervision Commission (NSC) and SCs also curbs "trial-centeredness" reform and centralization efforts (Meng 2021). One study suggests that, within the PRC's political-legal system, judicial reform needs to enhance judges' professionalism—without coming into conflict with that system—and it must address efficiency concerns (Sun and Fu 2022). Incorporating technology offers the potential to fulfil these requirements.

Amidst aspiration and dissension: exploring the impact of smart courts on judicial reform in contemporary legal scholarship

Smart courts and judicial reform are viewed as interconnected elements in modernizing the judicial system, with the People's Court Daily likening them to "two wheels of a vehicle and dual wings of a bird" (People's Court Daily 2020). The *Opinions of the Supreme People's Court on Accelerating the Construction of Smart Courts* also emphasizes that smart courts are intended to promote judicial reform through modern information technology (SPC 2017). It can be inferred that the SPC perceives smart courts as valuable tools in existing reform efforts, a notion endorsed by numerous scholars in the field. Li argues that the design of judicial reform shapes the goals of smart courts, using technology to enhance and showcase the achievements of reform, thereby demonstrating its progress in a data-driven manner (Li 2021).

Nonetheless, a conspicuous difference is apparent between the anticipated impact and the actual outcomes from smart courts. When assigning expectations to them, whether by the SPC or positive commentators, it appears that the focus is primarily on the issues arising from existing reforms within the judicial system, with no consideration given to external conditions, social structures, or China's political-legal system.

Although various scholars have offered critiques of smart courts' influence on judicial accountability reform, a definitive understanding is yet to emerge. Concerns have been raised regarding how smart courts may hinder accountability reform. Existing research highlights the possibility that in the event of errors judges could potentially evade or mitigate their liability by attributing responsibility to the computer system. This stems from an apparent distrust of judges and a belief that AI may be used to circumvent genuine accountability reform efforts (Gao 2022; Zhang 2022; Wang 2022; Ma 2018; Zheng 2020; Gao 2019). One researcher acknowledges that smart courts may inadvertently encourage the collectivization of responsibility.

Nevertheless, these systems also offer a clear record of judges' actions, facilitating the determination of individual responsibilities (Papagianneas 2023).

The following section analyzes four cases before a discussion takes place of the actual ramifications of smart courts for ongoing reform.

Smart Courts: A Digital Cage

This section analyzes data from the *Annual Reports*, with each of the four selected reports corresponding to a case. The four data-driven systems are: the Jiangsu Alert Platform for Inconsistent Judgements in Similar Cases (Wang et al. 2018); the evidence review mechanism in Guizhou (IT Department of Guizhou High People's Court 2018); the Automated Courtroom Inspection System in Hebei (Li 2019); and the Integrated Platform for Supervision and Constraint of Judicial Authority, which is also operational in Hebei (Research Team of the Integrated Platform at Hebei High People's Court 2022). These cases reveal that smart courts adopt a panoramic approach to the supervision of judges and other staff. Data-driven systems enable comprehensive collection of judicial behavior and case data, facilitating exhaustive and real-time judicial supervision. By converting case trial rules into algorithmic rules and employing automatic large-scale review and analysis, potential risks can be continuously identified, while normalizating judicial supervision.

Alert platform for inconsistent judgements in similar cases

A problem in Chinese courts is that similar cases are often decided differently (Ahl 2014). Since the start of the judicial reform process, increased judicial autonomy has led to differing legal standards being applied, even among judges sharing an office (He 2021). It is seen as a significant cause for public discontent with the judicial system since it creates a perception that court decisions are arbitrary and unfair (Ahl 2014). The report points out that the fairness of judicial decisions often comes from the consistency of the decisions, with consistency seen as the "bottom line of judicial fairness". Technology has ushered in a myriad of novel prospects for augmenting consistency.

The SPC Judicial Big Data Research Base, established by Jiangsu High People's Court and Southeast University, has developed the Alert Platform designed to promote consistency, that is, to prevent similar cases being decided differently. The platform aids judges with sentencing, case recommendations, and legal knowledge, while monitoring deviations. Currently operational in Jiangsu Province where 350 judges operate, the platform focuses on criminal cases, with plans to expand it to civil and administrative cases.

Jiangsu High People's Court addresses the problem of dissimilar decisions in similar cases by using the Alert Platform. It integrates four functions designed to prevent high levels of deviation, one that seeks to improve judges' writing-up of judgements, one for assisting judges during mediation, and one that lies beyond the scope of this study. Table 4.1 presents a concise summary of these functions.

Table 4.1 Summary of alert platform functions

Function	Description
Similar Case Recommendation	recommending cases similar to the one being considered by judges
Legal Knowledge Push	providing professional knowledge, legal provisions, and regulations
Intelligent Sentencing Support	analyzing laws, policies, guidelines, and historical sentencing patterns to offer judges recommended sentencing outcomes
Sentencing Deviation Alert	comparing the judge's sentencing decision with predicted outcomes; alerting if the deviation exceeds a threshold
Intelligent Error Correction	identifying text and delivery errors, including unprofessional language use, misspellings of amounts and units, etc.; detecting inadequately reasoned decisions

The Alert Platform could be of assistance to judges when it comes to increasing consistency of judgements and standardization of writing; however, these effects come from another role of the Alert Platform. It functions as a supervisor since it reviews and regulates judges' discretionary powers, which naturally results in consistency and standardization. The alert usually triggers supervision from the court presidents or chief judges. Judges are required to justify the difference if they insist on decisions with significant deviation, with the justification recorded and retained. If the decision cannot be justified, a similar judgement is mandatorily applied; otherwise, the Alert Platform considers them discrepancies and promptly notifies the court president (Zhang and Sun 2022). By implementing this process, the Alert Platform effectively ensures that the variations remain within predefined limits (Li 2021).

The Alert Platform has proven to be a highly effective tool in monitoring a vast number of judgments, addressing the limitations of the traditional random sampling inspection method, which fell short of achieving complete case coverage. With the Alert Platform, real-time monitoring of case deviations becomes possible, facilitating a comprehensive analysis of such deviations within the region. By accessing the big data analysis interface, court officials can swiftly grasp the overall picture and the development trends in case quality in the area.

Additionally, the Alert Platform streamlines court officials' management tasks, offering a more efficient alternative to the previous case-approval system. This targeted approach, known as "precise" management, only requires court officials to handle cases that the system signals as falling outside a certain range. Notably, this system operates as "silent" management, meaning court staff generally experience the presence of management only when there is a breach, rather than under circumstances of compliance or consistency.

In summary, the Alert Platform facilitates and strengthens the supervision of judges. In doing so, it contributes to consistency and enhances the standardization of the judicial process using a pre-designed programme, and yet it diminishes judges' autonomy by forcing judges to follow the algorithm.

Evidence review mechanism

The Guizhou High People's Court identified two enduring issues prevalent among the criminal cases remanded for retrial due to insufficient evidence in its Province. The first issue relates to insufficient evidence and flawed evidence-taking procedures. Evidence collection is the main task at the criminal investigation stage. Non-standardization of evidence collection has always been a problem in China, however. Investigating agencies primarily rely on suspect confessions, strengthening subjective judgment and overlooking case doubts (Wang and Tang 2021; Mu 2016). Investigating agencies selectively favor incriminating evidence while neglecting exculpatory evidence (Wang and Tang 2021; He 2014). Moreover, there is pervasive occurrence of torture-induced confessions which account for nearly all wrongful convictions, according to empirical studies (Wang and Tang 2021; Chen 2005). The second concern pertains to the lack of standardized evidence criteria among police departments, procuratorates, and courts. This issue arises due to the extensive autonomy of various agencies in their collection and utilization of evidence (Wang 2022).

On April 27, 2016, the Guizhou High People's Court, together with the Provincial People's Procuratorate and the Provincial Public Security Department, jointly formulated *Basic Evidence Requirements for Criminal Cases*, designed to serve as a fundamental reference point for automated evidence review (Guizhou High People's Court et al. 2016).

The evidence guidelines were subsequently integrated into a comprehensive case-handling system. Through accessing the system's operational interface and utilizing a designated button, the staff can gain access to a visual representation of the evidence in diagram form. The use of colors and symbols facilitates differentiation, while judges can conveniently access all material electronically. Moreover, the system automatically identifies any missing essential evidence relevant to the case.

The High People's Court of Guizhou Province attaches great importance to communication with public security organizations and procuratorates. They believe that evidence rules in court should be shared with public security organizations and procuratorates for the purpose of guiding investigations and prosecutions according to legal standards and promoting "trial-centeredness" criminal procedure reform. In response to the aforementioned challenges, the courts in Guizhou have implemented an evidence review mechanism encompassing criminal, civil, and administrative cases. This mechanism focuses solely on the processes related to criminal cases. For criminal cases, it is jointly utilized by the police department, the procuratorates, and the courts (Wang and Tang 2021). The evidence review mechanism comprises three key functions, as shown in Table 4.2. It also provides similar case recommendations, sentencing assistance, and deviation calculations. As these aspects have been previously discussed, further elaboration will not be provided here.

Table 4.2 Summary of evidence review function

Function	Description
Evidence Rules Guidance	prescribing norms for evidence collection and utilization in investigation departments
Evidence Check	using evidence rules to review evidential materials; alerting or rejecting cases lacking essential evidence
Evidence Analysis	identifying internal connections through logic mining; detecting contradictions between testimonies and evidence

Similar to the Alert Platform, the evidence review mechanism has proven itself beneficial to judges as well as staff at the procuratorate and in the police department. This mechanism significantly contributes to the standardization of evidence-related procedures (Liu and Chen 2019). Moreover, it serves as a supervision method that addresses concerns regarding the lack of professionalism among investigators. At each stage of the case, it offers reminders and facilitates real-time supervision, informing consistent collection and utilization of evidence (Wang 2021).

It endeavors to compel the police and procuratorate departments to adhere to the courts' rules during criminal procedures, striving in this way to elevate the courts' prominence in criminal proceedings. Its design integrates a unified evidentiary standard into the case management systems of the police, procuratorate, and courts, connecting the case platforms of various organizations involved in the criminal proceedings. It forces organizations to strictly implement the same rules of evidence at different stages of the proceedings. The report metaphorically describes this mechanism as a "filter" because it automatically sifts out cases with strong evidentiary value, that is, those with the required evidence for the trial phase, while intercepting cases with significant flaws and insufficient evidence during the investigation and prosecution stages. This "filter" therefore has the potential to foster a consistent standard among the "triangle" departments.

In summary, the evidence review mechanism facilitates the standardization of judicial procedures through a comprehensive supervision process. It ensures the enforcement of consistent evidence rules across the police department, procuratorate, and the court.

Automated courtroom inspection system

Research has highlighted the issue of judges' lack of professionalism during court hearings, evidenced by non-standardized courtroom language usage and inappropriate behaviors (Shao 2016; Liu 2022). The perceived deficiency in professional ethics among Chinese judges has drawn frequent criticism, given that professional ethics has always been viewed as an essential and integral aspect of judges' professionalism (Wang 2007).

In 2013, the SPC decided that courts should conduct simultaneous audio and video recordings of entire court hearings, "recording every trial (每庭必录)" (SPC 2013). In the following year, the SPC further proposed enhancing judicial transparency through public access to court hearings. Prior to this, the Hebei High People's Court had identified instances of judges ignoring court hearing rules and engaging in inappropriate conduct, including absenteeism, lateness, early or unauthorized departures, answering of phone calls, and dressing in non-standard attire. Such behaviors, if publicly exposed during the trial, had the potential to damage the image of judges, and their reputations.

Prior to the invention of the automatic court inspection system, court hearings were supervised via manual methods involving courtroom observation, surveillance footage monitoring, and live streaming; however, due to the restricted number of trial management staff, these manual approaches had limitations that made it impractical to continuously monitor every court hearing. Geographical distances, video-uploading processes, and the sheer volume of court hearings made it challenging for higher-level courts to maintain continuous monitoring, leading them to observe only the *Potemkin Village*. It became evident that relying solely on human resources would not suffice for a more thorough oversight.

To achieve more comprehensive monitoring, the Hebei High People's Court took an information-based and intelligent approach. By leveraging video image recognition and audio recognition technology, they combined the practices of "recording every trial" and "checking every trial." This approach enables the automated monitoring of live trial broadcasts or recordings aimed at standardizing judges' courtroom behaviors, strengthening the supervision from upper-level courts, and ultimately enhancing the image and authority of the judiciary.

The system monitors all court hearings throughout Hebei province, having adopted a hierarchical deployment model, with two levels of platforms established, one in the High People's Court and the other in all Intermediate People's Courts. As Table 4.3 shows, this system has two main functions.

Table 4.3 Summary of automated court inspection system

Function	Description
Checking Procedural compliance	Verifying court hearing recordings and timeliness; ensuring complete court transcripts and adherence to standardized processes.
Enforcing Courtroom Disciplines	Monitoring judges' behaviors and speech and offering real-time feedback to court officials; tracking behaviors like absence, lateness, phone use, etc.

The data gathered from the system's daily inspections, which undergo manual review, is accessible to court officials, judicial management departments, supervision departments, and higher-level courts, and is linked to judges' performance appraisals. When the system detects any questionable behavior or circumstance, it automatically captures evidence like screenshots and video recordings, which are then sent to relevant staff or departments with supervisory authority. The Hebei High People's Court's system allows for random checks of intermediate and primary courts, with intermediate courts in Hebei having the ability to supervise both intermediate and primary courts under their jurisdiction. The system also automatically scores courtroom trials based on specific rules. The evidence and scoring results feed into judges' performance appraisals.

The automatic courtroom inspection system significantly strengthens the supervision and management work of the administrative department, while enhancing the monitoring of efficiency and increasing pressure on judges to maintain professionalism. It ensures that every court hearing is assessed, considerably reducing the workload of supervision personnel. Through real-time checks during court hearings and subsequent inspections, the system achieves comprehensive and continuous court supervision. With a single back-end administrator, the system intelligently inspects all courtroom hearings, autonomously capturing screenshots, recording videos, and generating inspection logs that identify instances of non-compliance, effectively compelling judges to uphold courtroom discipline.

In China, the principle of transparency is imbued with strong propagandist and educational implications; it is a principle primarily aimed at bolstering the legitimacy of the ruling regime. It involves presenting a carefully curated image of the judiciary while simultaneously shielding the public from any trials that could potentially jeopardize the interests of the Party-state (Ahl and Sprick 2017). Courts in Hebei have been live-streaming some court hearings. Where cases have been streamed, the public can also replay the recordings (SPC 2014). This live-streaming has garnered considerable attention from the public (SPC 2011). Through the publication of court hearing videos where judges are "well-behaved", it is possible for this transparency to foster an image of a professional judiciary, helping to build public trust, which ultimately contributes to the Party-State's legitimacy.

To summarize, courts in Hebei have automated court hearing supervision to enforce procedural compliance and regulate judges' behaviors. It facilitates supervision both within one court and between higher-level courts and lower-level courts, increasing pressure on judges to exhibit professionalism. The effect of the policy on judicial transparency, meanwhile, is to improve the image of the courts and have a positive effect on the regime's legitimacy.

Integrated platform for supervision and constraint of judicial authority

Corruption has always been a serious issue in Chinese courts (Li 2014). The establishment of NSC and SCs was intended partly as an anti-corruption mechanism, reflecting the party-state's desire to strengthen and broaden its self-monitoring capacities. The SCs are exclusively responsible for investigating corruption cases, supplanting the procuratorates' jurisdiction. The SCs at both the central and local levels "work in joint offices(合署办公)" with the CDIs at each administrative level, which is actually "one institution, two labels (一个机构, 两块牌子)". Prior to this reform, the CDIs could launch disciplinary investigations only against party members in public posts. Now, the CDIs are entitled to question and investigate both CCP members and non-party members holding public positions (Meng 2021). To ensure a more comprehensive oversight, the SCs and CDIs often establish embedded offices within other entities, for example, courts. These offices, working as extensions of SCs and CDIs, concentrate on upholding Party loyalty, principles, integrity, and compliance with regulations. They are also instructed to pay particular attention to the leadership teams and key personnel within the host institution (CCPCC 2022). Scholars have noticed a trend in SCs and CDIs expanding their activities beyond routine supervision of disciplinary violations which are not emphasized and regularized (Laha 2019). Modern technology enables upgrades and extensions of supervision, strengthening the exercise of judicial authority.

Some studies reveal that this anti-corruption reform has weakened the powers of both the procuratorates and the courts in criminal procedures vis-à-vis the SCs, left openings for local political interests to intervene, and extended the party's power into the judicial realm (Meng 2021; Tan and Wang 2020). In corruption cases, the extensive investigative powers possessed by SCs and CDIs could deepen the "investigation-centeredness" approach which goes against the "trial-centeredness" criminal proceedings reform (Tan and Wang 2020).

Against this backdrop, Hebei High People's Court has pointed out some persistent issues in its courts, such as delays in filing cases, prolonged unresolved cases, and unauthorized property seizures. In response to these challenges, the Court has made the decision to establish a platform that allows for inquiry into and tracing of power operations. It established the "Integrated Platform for Supervision and Constraint of Judicial Authority (hereafter referred to as the Integrated Platform)" with its dispatched organization (hereby referred to as the DO).

The Integrated Platform comprises two distinct sub-platforms. The first one is called the "Automated Supervision and Early Warning Platform at Hebei Courts" (referred to as the Court Platform). The second one is known as the "Smart Supervision Platform for the Dispatched Organisation " (referred to as the DO Platform). As the names imply, the court system utilizes the former, while the DO adopts the latter.

The Court Platform identifies and presents key cases, events, and "nodes" to court officials for review. This includes a range of cases like long-pending, flawed, politically sensitive, organized crime, and serious offense cases. It encompasses 12 nodes, monitoring seven trial-related nodes (such as unauthorized extension or suspension) and five enforcement-related nodes (for example, overdue network search). The platform also oversees "high-risk" and "medium-risk" judges[24] and disciplines individuals. It autonomously identifies supervisory matters based on performance requirements and escalates issues when handling time exceeds limits, thus establishing a multi-layered supervision mechanism. The entire process remains traceable and reviewable.

The DO Platform, akin to the Court Platform, also focuses on key cases, events, nodes, and personnel; however, its distinguishing feature is the implementation of a more elevated multi-layered supervision mechanism, warranting particular attention. Upon identifying issues, the DO Platform sends a "Prompt Action Card (督办卡)", demanding timely rectification and resolution. Reminders are issued to prompt court officials to fulfil their supervision and management responsibilities, strengthening the accountability of court leaders. In instances where court leaders fail to address the identified concerns within a given timeframe, the DO Platform notifies the DO, leading to the reinforcement of their accountability through

[24] The report gave no explanation for what constitutes "high-risk" or "medium-risk", but in this context, it seems to refer to integrity risk.

methods such as “yue tan (约谈)”,[25] and reducing any reluctance among court leaders to fulfil their supervision duties.

The Integrated Platform enforces multi-layered supervision, with judges at the lowest level and the DO at the apex, promptly activating the warning process when issues arise. Unresolved problems escalate to higher authorities, leading to accountability advice and performance assessment for those showing poor performance. Various measures, like withholding bonuses, deferring promotions, and withdrawing quotas, are applied to discourage poor performance.

This study contends that the implementation of the Integrated Platform has resulted in strengthened supervision of the court by the DO, signifying a further consolidation of the Party’s control over the judicial system. This finding expands upon He’s research, which highlighted the formidable control exerted by the Party through the SCs (He 2021). Through the Integrated Platform, the DO ensures the integrity of judges and monitors the job performance of court staff in a timely, standardized, and effective manner, establishing a multi-layered supervision mechanism that reinforces the Party’s dominant authority. It can even be argued that, in this context, the courts are becoming increasingly subordinate to the DO.

The DO’s effect is consolidated via the Integrated Platform. Courts experience further diminution of their autonomy in comparison to the SCs, affecting not only criminal procedures but also extending to other domains like civil procedures because this multi-layered supervision is applied in all case proceedings. The “investigation-centeredness” feature is also reinforced, given the augmented authority granted to the SCs and CDIs by the Integrated Platform during the investigation process.

The “legitimate” (He 2021) control exerted by the Party ensures a measure of accountability with the implementation of the Integrated Platform, as it ensures comprehensive recording and traceability of all supervisory operations carried out by the DO. While the Party relies on its agents, the DOs, for effective control over the courts, it also shows caution in requiring accountability measures. It potentially limits illegitimate local political interventions, supporting SPC’s centralization efforts. It is important to acknowledge, however, that ultimate control over the courts’ centralization remains with the Party.

In conclusion, the Integrated Platform tightens the Party’s control over the courts. The multi-layered supervision mechanism makes courts more subordinate, and extends investigation-centric approaches. It introduces accountability for local party agencies, potentially reducing illegitimate local interventions and bolstering the SPC’s centralization ambition, but always under the Party’s ultimate control.

[25] In China, “约谈” (yue tan) refers to a formal meeting or conversation between higher-level authorities and lower-level officials or individuals, often used to discuss specific issues, offer guidance, and convey expectations or concerns.

Data-Driven Promise: A Detriment to Judicial Reform

Ji argues that addressing judge-related issues such as misjudgments, arbitrariness, and corruption in the Chinese legal system through improvements in judge selection, qualifications, education, and accountability is more effective than relying on computer sentencing (Ji et al. 2016). I further posit that smart courts augment the professionalism of judges by enforcing efficiency, consistency, and procedural compliance, as well as addressing malfeasance, corruption, illegitimate intervention, and dereliction of duty. With this enhancement, smart courts may improve the image of the courts and contribute to the regime's legitimacy. Although they may also "treat the symptom, not the disease", smart courts serve as a conduit for delivering efficient solutions and elevating professionalism without encroaching upon the political system.

The provincial pilots discussed above are an attempt at soft-centralization through technology. They facilitate the ability of higher-level courts to supervise lower-level courts and their judges. Smart courts reconfigure the control relationship between judges and court officials. Simultaneously, the pilots reinforce and confirm the previously administrative relationship between upper-level and lower-level courts. Supervision and control within the court system becomes intangible but more direct and effective. The monitoring structure that had already been broken is reconstructed, even surpassing the previous level of control. In this sense, the court system is further centralized, as suggested in the findings of Stern et al. (2021). We could even argue that primary people's courts are becoming circuit courts of the high people's courts since the provincial pilots establish consistent standards as well as empower high people's courts to oversee courts across the entire province.

Current judicial reform focuses on granting judges more autonomy, however. Smart courts reject the delivery of autonomy through re-establishing control over judges via intangible data-driven systems rather than tangible organizational structures.

Smart courts take the opposite course from "trial-centeredness" reform since the data-driven systems cannot change but rather enhance the "assembly line". The Guizhou case highlights how investigations, prosecutions, and court decisions are propelled by a shared algorithm, with each case's outcome predetermined upon entry into the system. As a result, smart courts strengthen the original relationship between police departments, procuratorates, and courts, hindering the "trial-centeredness" reform.

According to the SPC, all responsibility falls on judges when utilizing AI in their decisions (SPC 2022). This means that judges are expected to assume responsibility even if errors occur due to the machine's actions; consequently, judges face a dilemma—disobeying the machine may result in negative assessments and heightened scrutiny, while adhering to its decisions makes them accountable for any potential errors.

As stated above, smart courts may address some issues in judicial reform and perform their anticipated functions. Contrary to expectations, however, this study reveals that smart courts do not yield substantial benefits but instead have a detrimental impact on judicial reform. The Party's authority remains supreme. Smart courts might have failed to promote judicial reform, but they have successfully tightened the party-state's power over the courts and judges.

Conclusion

This chapter set out to determine the implications of the operations of smart courts for judicial reform. The post-2013 reforms pursued goals associated with judges' professionalism, legal procedures, autonomous decision-making, accountability, and centralization of the judiciary, but these reforms have encountered numerous challenges, and the reform journey will continue to be demanding. Smart courts are envisioned as complementary to these judicial reforms, aiming to facilitate their realization and achieve their objectives. Through the investigation of four cases, this study provides nuanced insights into the multifaceted impact of smart courts on judicial reform. While smart courts were initially anticipated to act as a remedy for specific issues and to serve as a catalyst for enhancing professionalism, the findings in this study indicate that smart courts do not yield substantial benefits and, indeed, are detrimental to the trajectory of judicial reform.

While smart courts may contribute to impoving the professionalism of judges by fostering efficiency, consistency, procedural compliance, and addressing instances of misconduct, thereby potentially elevating the public perception of courts and augmenting the regime's legitimacy, their impact remains superficial, with core structural challenges within the prevailing political system being avoided. Moreover, through their implementation, smart courts recalibrate the equilibrium of control between judges and court officials, while concurrently reinforcing the administrative dynamics between higher and lower-level courts, ultimately leading to further centralization of the judicial system. Notably, smart courts contradict the emphasis on judges' autonomy and "trial-centeredness" in the current reform agenda. The data-driven systems also cause judges to encounter an accountability dilemma.

Smart courts, therefore, despite appearing to assist in judicial reform, actually present yet another obstacle to the already challenging reform process. They have the potential to hinder, distort, or even undermine the current efforts aimed at reforming the judiciary. Simultaneously, however, smart courts undeniably serve as an instrument for reinforcing the Party's authority, consolidating the party-state's control over the judiciary and judicial actors.

References

Ahl, Björn. 2014. "Retaining Judicial Professionalism: The New Guiding Cases Mechanism of the Supreme People's Court." *The China Quarterly* 217:121–39. https://doi.org/10.1017/S0305741013001471.

Ahl, Björn. 2021. "Post-2013 Reforms of the Chinese Courts and Criminal Procedure." In *Chinese Courts and Criminal Procedure*, 1–28. Cambridge University Press.

Ahl, Björn, and Daniel Sprick. 2017. "Towards Judicial Transparency in China: The New Public Access Database for Court Decisions." *China Information* 32 (1): 3–22. https://doi.org/10.1177/0920203X17744544.

Biddulph, Sarah, Elisa Nesossi, and Susan Trevaskes. 2017. "Criminal Justice Reform in the Xi Jinping Era." *China Law and Society Review* 2 (1): 63–128. https://doi.org/10.1163/25427466-00201002.

CCPCC (CCP Central Committee). 2014. Zhonggong zhongyang guanyu quanmian tuijin yifazhiguo zhongdawenti de jueding中共中央关于全面推进依法治国若干重大问题的决定[The Communist Party of China Central Committee's Decision on Major Issues Concerning the Comprehensive Advancement of the Rule of Law in China]. Accessed July 23, 2023. https://www.gov.cn/zhengce/2014-10/28/content_2771946.htm

CCPCC. 2022. Jijian jiancha jiguan paizhu jigou gongzuo guize纪检监察机关派驻机构工作规则[Rules for the Work of Disciplinary Inspection and Supervision Dispatched Organisations]. Accessed July 23, 2023. https://www.gov.cn/zhengce/2022-06/28/content_5698233.htm

Chan, Peter. 2015. "An Uphill Battle: How China's Obsession with Social Stability Is Blocking Judicial Reform." *Judicature* 100 (3).

Chen, Guangzhong陈光中, and Long, Zongzhi 龙宗智.2013.Guanyu Shenhua sifa gaige ruogan weiti de sikao关于深化司法改革若干问题的思考[Reflections on Several Issues Regarding Deepening Judicial Reform]. Zhongguo faxue中国法学(04):5–14.

Chen, Weidong陈卫东.2016.Yi shenpan wei zhongxin: Dangdai zhongguo xingshi sifa gaige de jidian 以审判为中心:当代中国刑事司法改革的基点[Putting Adjudication at the Center: The Foundation of Contemporary Criminal Judicial Reform in China]. Faxuejia法学家(04):1-15+175.

Chen, Weidong陈卫东.2017.Sifa zerenzhi gaige yanjiu司法责任制改革研究[Research on Judicial Accountability System Reform]. Faxue zazhi法学杂志(08):31–41.

Chen, Xingliang陈兴良.2005.Cuoan heyi xingcheng错案何以形成[The Formation of Wrongful Cases]. Zhejiang jingcha xueyuan xuebao浙江警察学院学报(05): 12–13.

Chen, Yongsheng陈永生, and Bai, Bing 白冰.2016. Faguan, jianchaguan yuanezhi gaige de xiandu法官、检察官员额制改革的限度[The Limits of Reforms to the Judge and Procurator Quota System]. Bijiaofa yanjiu比较法研究(02):21–48.

Cheng, Jinhua程金华, and Li, Xueyao 李学尧.2012.Falv bianqian de jiegouxing zhiyue – Guojia, shichang yu shehui hudong zhongde zhongguo lvshi zhiye法律变迁的结构性制约——国家、市场与社会互动中的中国律师职业[The Structural Constraints of Legal Transformation: The Chinese Legal Profession in the Interaction of State, Market, and Society]. Zhongguo shehui kexue中国社会科学(07):101-122+205.

Deng, Kai邓凯.2021. Shilun zhongguo sifa xinxihua de yanjin jiqi jishu lujing试论中国司法信息化的演进及其技术路径[A Discussion on the Evolution of Judicial Informatization in China and its Technological Path]. Juece tansuo (xia)决策探索(下)(01):49–53.

Gao, Tongfei高童非.2019.Woguo xingshi sifa zhidu zhongde xieze jizhi – yi fayuan he faguan weizhongxin我国刑事司法制度中的卸责机制——以法院和法官为中心[The Exemption Mechanism in China's Criminal Justice System: Focusing on Courts and Judges]. 浙江工商大学学报(05),102–119. doi:10.14134/j.cnki.cn33-1337/c.2019.05.010.

Gao, Tongfei高童非.2022. Jingti "yian tongpan" – Leian caipan jizhi de gongneng yuewei yu guiwei 警惕"异案同判"——类案裁判机制的功能越位与归位[Be Cautious of "Uniform Judgments for Divergent Cases": The Overreach and Restoration of the Mechanism of Similar Case Adjudication]. Nantong daxuebao(shehui kexueban)南通大学学报(社会科学版)(01):101–110.

Greitens Sheena Chestnut. 2020. The Saohei Campaign, Protection Umbrellas, and China's Changing Political-Legal Apparatus. Accessed July 23, 2023. https://www.prcleader.org/greitens-1

Guizhou High People's Court, Guizhou Provincial People's Procuratorate, and Guizhou Provincial Public Security Department. 2016. Guizhousheng xingshi anjian jiben zhengju yaoqiu贵州省刑事案件基本证据要求[The Basic Requirements for Criminal Case Evidence in Guizhou Province]. Accessed July 23, 2023. http://www.drxsfd.com/xf/xx2_sj.asp?bh=1516

Guo, Zhiyuan. 2021. "Live Witness Testimony in the Chinese Criminal Courts." In *Chinese Courts and Criminal Procedure*,183–207. Cambridge University Press.

He, Jiahong何家弘.2014.Dangjin woguo xingshi sifa de shida wuqu当今我国刑事司法的十大误区[The Top Ten Misunderstandings in China's Current Criminal Justice System]. Qinghua faxue清华法学(02):47–67.

He, Xin. 2021. "Pressures on Chinese Judges Under Xi." *SSRN Journal*. https://doi.org/10.2139/ssrn.3761675.

IT Department of Guizhou High People's Court.2018. Sifa zhengju de dashuju fenxi – yi guizhou fayuan de Shijian weili司法证据的大数据分析———以贵州法院的实践为例[Big Data Analysis of Judicial Evidence: A Case Study of Guizhou Courts' Practice]. In Annual Report on Informatization Of Chinese Courts No.2, 364–380. Shehui kexue wenxian chubanshe.

Ji, Weidong et al.季卫东等. 2016. Zhongguo de sifa gaige中国的司法改革[Judicial Reform in China]. Beijing: Falv chubanshe.

Ji, Weidong, and Xi Lin. 2022. *Towards the Rule of Law in China:* Cambridge University Press.

Jiang, Feng姜峰.2016.Yangdi guanxi shijiao xiade sifa gaige: Dongli yu tiaozhan央地关系视角下的司法改革:动力与挑战[Judicial Reform from the Perspective of Central-Local Relations: Motivation and Challenges]. Zhongguo faxue中国法学(04):127–142.

Laha, Michael. 2019. The National Supervision Commission: From "Punishing the Few" toward "Managing the Many". Accessed July 23, 2023. https://www.ccpwatch.org/single-post/2019/07/15/the-national-supervision-commission-from-punishing-the-few-toward-managing-the-many

Li, Anthony H. F. 2016. "Centralisation of Power in the Pursuit of Law-Based Governance." *chinaperspectives* 2016 (2): 63–68. https://doi.org/10.4000/chinaperspectives.6995.

Li, Jianli李建立. 2019. Hebei fayuan tingshen zidong concha xintong jianshe diaoyan baogao河北法院庭审自动巡查系统建设调研报告[Research Report on the Construction of Hebei Court's Automated Courtroom Inspection System]. In *Annual Report on Informatization Of Chinese Courts No.3*, 213–226.

Li, Wenwen李文文. 2021. Zhihui shenpan ruhe xiaojian chabie daiyu智慧审判如何消减差别待遇[How Smart Adjudication Reduces Differential Treatment]. Pingwei·Jingdian品位·经典(08):121–124.

Li, Xin李鑫. 2021. Zhihui fayuan jianshe de lilun jichu yu zhongguo shijian智慧法院建设的理论基础与中国实践[The Theoretical Foundation and Chinese Practice of Smart Court Construction]. Zhengfa luncong政法论丛(05):128–138.

Li, Yuwen. 2014. *The Judicial System and Reform in Post-Mao China: Stumbling Towards Justice*. Farnham, Surrey: Ashgate.

Lieberthal, Kenneth G., and David M. Lampton. 1992. *Bureaucracy, Politics, and Decision Making in Post-Mao China.* Berkeley: University of California Press

Liebman, Benjamin L. 2007. “China’s Courts: Restricted Reform.” *The China Quarterly* 191:620–38. https://doi.org/10.1017/S0305741007001610.

Liu, Pinxi刘品新, and Li Chen 陈丽. 2019. Shujuhua de tongyi zhengju biaozhun数据化的统一证据标准 [Unified Standards of Data in Evidence]. Guojia jianchaguan xueyuan xuebao国家检察官学院学报(02):129–143.

Liu, Yanhong刘艳红. 2019. Dashuju shidai shenpan tixi he shenpan nengli xiandaihua de lilun jichu yu Shijian zhankai大数据时代审判体系和审判能力现代化的理论基础与实践展开[The Theoretical Foundation and Practical Development of Modernizing the Judicial System and Judicial Capability in the Big Data Era]. Anhui daxue xuebao(zhexue shehui kexueban)安徽大学学报(哲学社会科学版)(03):96–107.

Liu, Ziyue刘紫月. 2022 .Xingshi tingshen Zhong tingshen jilv kongzhi cunzaide wenti yu duice刑事庭审中庭审纪律控制存在的问题与对策[The Problems and Countermeasures of Trial Discipline Control in Criminal Trials]. Fazhi bolan法制博览(03):157–159.

Long, Zongzhi龙宗智. 2015. “Yi shenpan wei zhongxin” de gaige jiqi xiandu“以审判为中心”的改革及其限度[The Reform of “Putting Trial at the Centre” and Its Limitations]. Zhongwai faxue中外法学(04):846–860.

Lubman, Stanley B. 2000. *Bird in a Cage: Legal Reform in China After Mao.* Great Britain: Stanford University Press.

Ma, Changshan马长山. 2018. Rengong zhinengde shehui fengxian jiqi falv guizhi人工智能的社会风险及其法律规制[The Social Risks of Artificial Intelligence and Its Legal Regulations]. Falv kexue(xibei zhengfa daxue xuebao)法律科学(西北政法大学学报)(06):47–55.

Meng, Ye. 2021. “Judicial (Dis)Empowerment and Centralisation Efforts.” In *Chinese Courts and Criminal Procedure*, 109–43. Cambridge University Press.

Meng, Ye. 2023. “The Limits of Judicial Reforms: How and Why China Failed to Centralize Its Court System.” *The China Quarterly*, 1–15. https://doi.org/10.1017/S0305741023000358.

Mertha, Andrew C. 2005. “China’s “Soft” Centralization: Shifting Tiao/Kuai Authority Relations.” *The China Quarterly* 184:791–810. https://doi.org/10.1017/S0305741005000500.

Mu, Shuqin穆书芹. 2016. Zhencha jieduan xingshi cuoan fangfan zhi zhencha linian, xingweiyu zhidu goujian侦查阶段刑事错案防范之侦查理念、行为与制度构建[Preventing Wrongful Criminal Cases during the Investigation Stage: The Construction of Investigative Concepts, Practices, and Institutions]. Zhongguo xingshifa zazhi中国刑事法杂志(01):89–101.

Ng, K., & X He. 2017. *Embedded Courts: Judicial Decision-Making in China.* Cambridge: Cambridge University Press.

Ng, Kwai Hang and Xin He. 2021. “Blood Money and Negotiated Justice in China.” In *Chinese Courts and Criminal Procedure*, 208–34. Cambridge University Press.

Papagianneas, Straton. 2022. “Towards Smarter and Fairer Justice? A Review of the Chinese Scholarship on Building Smart Courts and Automating Justice.” *Journal of Current Chinese Affairs* 51 (2): 327–47. https://doi.org/10.1177/18681026211021412.

Papagianneas, Straton. 2023. “Automating Intervention in Chinese Justice: Smart Courts and Supervision Reform.” *SSRN Journal.* https://doi.org/10.2139/ssrn.4383929.

People’s Court Daily. 2020. Jixu Shenhua sifa gaige, jiakuai jianshe zhihui fayuan.继续深化司法改革 加快建设智慧法院[Continue to Deepen Judicial Reform and Accelerate the Construction of Smart Courts]. Accessed July 23, 2023. http://rmfyb.chinacourt.org/paper/html/2020-11/06/content_173520.htm?div=-1

Research Team of the Integrated Platform at Hebei High People's Court. 2022. Hebei fayuan 'yitihua shenpanquan jiandu zhiyue pingtai'jianshe yingyong diaoyan baogao 河北法院"一体化审判权监督制约平台"建设应用调研报告[Hebei Courts' Research Report on the Construction and Application of the "Integrated Platform for Judicial Authority Supervision and Constraint"]. In *Annual Report on Informatization Of Chinese Courts No.6*, 250–266. Shehui kexue wenxian chubanshe.

Shao, Shanshan邵珊珊. 2016. Faguan tingshen yuyan guifanhua yanjiu – yi yuyan xingwei lilun wei fenxi shijiao法官庭审语言规范化研究——以言语行为理论为分析视角[Research on the Standardization of Judges' Trial Language: An Analysis Perspective Based on Speech Act Theory]. Fazhi yu shehui法制与社会(22):133–134.

Song, Yuansheng宋远升. 2017. Jingyinghua yu zhuanyehua de mishi – faguan yuanezhi de kunjing yu chulu精英化与专业化的迷失——法官员额制的困境与出路[The Dilemma and Way Out of Elitism and Professionalism Loss: The Predicament of the Judge Quota System]. Zhengfa luntan政法论坛(02):101–117.

SPC. 2011. Hebei langfang fayuan baichang wangshang tingshen zhibo cu yangguang sifa河北廊坊法院百场网上庭审直播促阳光司法[Hebei Langfang Court's Hundred Online Court Hearings Live Broadcasts Promote Transparent Judiciary]. Accessed July 23, 2023. https://www.court.gov.cn/zixun-xiangqing-11190.html

SPC. 2013. Guanyu tuijin sifa gongkai sanda pingtai jianshede ruogan yijian关于推进司法公开三大平台建设的若干意见[Several Opinions on Promoting the Construction of Three Major Platforms for Judicial Transparency]. Accessed July 23, 2023. https://splcgk.court.gov.cn/gzfwww/gkzn/gkznDetails?id=2c9030ab5fde9d87015fe69be2c60059

SPC. 2014. Hebei fayuan wangluo zhibo tingshen cu sifa gongping gongzheng河北法院网络直播庭审促司法公平公正[Hebei Courts Use Online Live Streaming of Court Hearings to Promote Judicial Fairness and Justice]. Accessed July 23, 2023. https://www.court.gov.cn/zixun-xiangqing-12928.html

SPC. 2016. Zhouqiang: Tuijin renmin fayuan xinxihua jianshe zhuanxing shengji周强：推进人民法院信息化建设转型升级[Zhou Qiang: Promote the Informationization Construction and Transformation Upgrade of the People's Courts]. Accessed July 23, 2023. http://www.cac.gov.cn/2016-02/01/c_1117956472.htm

SPC. 2017. Guanyu jiakuai jianshe zhihuifayuan de yijian关于加快建设智慧法院的意见[Opinions on Accelerating the Construction of Smart Courts]. Accessed July 23, 2023. http://gongbao.court.gov.cn/Details/5dec527431cdc22b72163b49fc0284.html

SPC. 2022. Guanyu guifan he jiaqiang rengongzhineng sifa yingyong de yijian关于规范和加强人工智能司法应用的意见[The Opinions on Regulating and Strengthening the Applications of Artificial Intelligence in the Judicial Fields]. Accessed July 23, 2023. https://www.court.gov.cn/zixun/xiangqing/382461.html

Stern, Rachel et al. 2021. "Automating Fairness? Artificial Intelligence in the Chinese Court.". Columbia Journal of Transnational Law (59): 515–553. Available at: https://scholarship.law.columbia.edu/faculty_scholarship/2940

Sun, Ying, and Hualing Fu. 2022. "Of Judge Quota and Judicial Autonomy: An Enduring Professionalization Project in China." *The China Quarterly* 251:866–87. https://doi.org/10.1017/S0305741022000248.

Tan, Shigui 谭世贵 and Jianlin Wang王建林.2020.Lun jijian jiancha zhidu gaige yu baozhangjianchaquan shenpanquan de yifa duli xingshi论纪检监察制度改革与保障检察权审判权的依法独立行使[Reform of the Disciplinary Inspection and Supervision System and Safeguarding the Independent Exercise of Procuratorial and Judicial Powers in Accordance with the Law]. Guizhou minzu daxue xuebao (zhexue shehui kexueban)贵州民族大学学报(哲学社会科学版)(02):189–208.

Wang, Chenguang王晨光. 2007. Faguan zhiyehua he faguan zhiyedaode jianshe法官职业化和法官职业道德建设[Judicial Professionalism and Judicial Professional Ethics Construction]. Jiangsu shehui kexue江苏社会科学(01):102–111.

Wang, Jingjing王晶晶. 2022. Rengong zhineng zai xingshi zhengju shencha panduan zhongde yingyong yanjiu人工智能在刑事证据审查判断中的应用研究[Research on the Application of Artificial Intelligence in Criminal Evidence Examination and Judgment]. Henan caijing zhengfa daxue xuebao河南财经政法大学学报(03):43–53.

Wang, Lusheng王禄生, Kun Liu 刘坤, Xiangyang Du 杜向阳, and Liang Yanyuan梁雁圆. 2018. Jiangsu fayuan "tongan butongpan yujing pingtai"diaoyan baogao江苏法院"同案不同判预警平台" 调研报告[Research Report on Jiangsu Court's "Alert Platform for Inconsistent Judgements in Similar Cases"]. In Annual Report on Informatization Of Chinese Courts No.2, 351–363. Shehui kexue wenxian chubanshe.

Wang, Ran王燃. 2021. Dashuju sifa jiandu jizhi yanjiu大数据司法监督机制研究[Research on the Mechanism of Big Data Judicial Supervision]. Hunan keji daxue xuebao湖南科技大学学报(社会科学版)(03):132–141.

Wang, Wenyu王文玉. 2022. Sifa rengong zhineng: Shixian caipan zhengyi de xinlujing司法人工智能：实现裁判正义的新路径[Judicial Artificial Intelligence: A New Path to Achieve Judicial Justice]. Dalian ligong daxue xuebao(shehui kexue ban)大连理工大学学报(社会科学版)(06):100–109.

Wang, Xiumei王秀梅, and Ling Tang 唐玲.2021.Rengong zhineng zai fangfan xingshi cuoan zhongde yingyong yu zhidu sheji人工智能在防范刑事错案中的应用与制度设计[The Application and Institutional Design of Artificial Intelligence in Preventing Wrongful Criminal Cases]. Faxue zazhi法学杂志(02):97–107.

Wang, Yaxin王亚新, and Qian Li 李谦.2014. Jiedu sifa gaige – Zouxiang quanneng, ziyuanyu Zeren zhixin de quanheng解读司法改革——走向权能、资源与责任之新的均衡[Interpreting Judicial Reform: Towards a New Balance of Power, Resources, and Responsibilities]. Qinghua faxue清华法学(05):103–113.

Wang, Yueduan. 2020. "Overcoming Embeddedness: How China's Judicial Accountability Reforms Make Its Judges More Autonomous." *Fordham International Law Journal* 43.

Wang, Yueduan. 2021. "'Detaching' Courts from Local Politics? Assessing the Judicial Centralization Reforms in China." *The China Quarterly* 246:545–64. https://doi.org/10.1017/S0305741020000740.

Wei, Xiaonan魏晓娜.2020.Yi shenpan wei zhongxin de susongzhidu gaige: Shixiao, pingjing yu chulu以审判为中心的诉讼制度改革:实效、瓶颈与出路[The Trial-Centered Procedural Reform: Effectiveness, Bottlenecks, and Solutions]. Zhengfa luntan政法论坛(02):155–172.

Zhang, Jianwei张建伟. 2015. Shenpan zhongxin zhuyi de shizhi neihan yu shixian tujing审判中心主义的实质内涵与实现途径[The Essence and Implementation Approaches of Trial-Centeredness]. Zhongwai faxue中外法学(04):861–878.

Zhang, Linghan张凌寒. 2022. Zhihui sifazhong jishu yilai de yinyou ji yingdui智慧司法中技术依赖的隐忧及应对[The Hidden Concerns and Responses to Technological Dependency in Smart Justice].Fazhi yu shehui fazhan 法制与社会发展(04):180–200.

Zhang, Shejun张社军, and Yaxuan Sun 孙亚轩. 2022. Tongan butongpan de leixinghua fenxi ji dashuju duice – Henan fayuan "leian zhiyin" xitong de xianzhuang ji gexin jinlu同案不同判的类型化分析及大数据对策——河南法院"类案指引"系统的现状及革新进路[Typological Analysis of Divergent Judgments in Similar Cases and Big Data Countermeasures: The Current Status and Innovative Path of the "Similar Case Guidance" System in Henan Courts]. Jingmao falv pinglun经贸法律评论(02):145–158.

Zheng, Xi郑曦. 2020. Rengong zhineng jishu zai sifa caipan zhongde yunyong ji guizhi人工智能技术在司法裁判中的运用及规制[The Application and Regulation of Artificial Intelligence Technology in Judicial Adjudication]. Zhongwai faxue中外法学(03):674–696.

CHAPTER 5

Balancing Control and Engagement: China's Sociotechnical Imaginary in Facial Recognition Technology

Xin Gu, Gavin Smith, Neil Selwyn, Mark Andrejevic, and Chris O'Neill

Abstract

This chapter explores the public response to facial recognition technology (FRT) in China and the complex interplay between state control and individual agency, reflexive of various social dynamics which influence how individuals reconcile the benefits of security and convenience with concerns over privacy and surveillance. We argue that its implementation extends China's sociotechnical imaginaries by offering individuals a mechanism to articulate their concerns and resistance at the grassroots level. Despite the government's oversight of FRT, the technology inadvertently opens channels for public dissent and negotiation. Through a case study of a micro sociocultural context of a middle-class housing compound in Shanghai, the chapter reveals how China's approach to automated surveillance technologies like FRT must continually reconfigure the delicate balance between social control and civic participation. It also considers how the pervasive use of FRT for mass surveillance may destabilize this fragile equilibrium, posing potential risks to the state's ability to maintain both governance and public trust.

Keywords: Sociotechnical imaginaries; Facial recognition technology; Surveillance; Biometrics; Chinese nationalism

Introduction

China is leading the world in the development and use of facial recognition technology (FRT hereafter). The country already has over 400 million CCTV cameras installed, making it the world's largest monitoring system (Bischoff 2023). Many of these cameras have built-in facial recognition capabilities as China develops a comprehensive strategy that prioritizes technological advancement, societal governance, and national security imperatives. The country has positioned itself as a world leader in the deployment and development of FRT technologies, redefining modern surveillance practices and repositioning the boundaries of individual privacy in the age of AI (Knight 2023; Davis 2021; Schmitz 2018). Much of the focus of media and academic coverage has been on the role played by FRT in the Chinese "social credit system," which uses identification with the stated ideal of serving national goals and priorities. This chapter draws on a case study of the local response to a proposed FRT system in a residential housing complex to explore some of the reaction "on the ground" to emerging uses of the technology.

As is the case elsewhere, facial recognition in China refers to the automatic processing of digital images containing individual faces for the purposes of verification, identification, and classification of these individuals (Shi 2022). Verification mainly refers to one-to-one facial authentication circumstances, such as at border control, or to an ID card photo being used to compare real-time face information to determine whether the person is the one represented by the ID card. Identification refers to one-to-many face comparisons, which has sparked controversy worldwide. China is among the very few countries that actively utilize one-to-many identification for purposes such as law enforcement. The Skynet system ("tianwang" or "king of heaven") run by law enforcement for the purpose of identifying criminals is one such example (Chen and Li 2017). Classification refers to various facial analyses such as calculating and profiling facial features to determine a person's gender, age, race and other characteristics. FRT has emerged as a powerful tool with diverse applications in China, ranging from security and surveillance to authentication, and even to leisure and entertainment. Apart from the deployment of FRT in combating crime and terrorism, many of the capabilities of FRT have not yet been widely adopted outside China (Sudworth 2017). These include:

- Surveillance: FRT has been deployed in a vast network of surveillance CCTV cameras in public spaces, transport systems, hospitals, and other public security systems in major cities. They allow one-to-many tracking and one-to-one identification of individuals. Data collected can be used by law enforcement for maintaining public safety and security.
- Education: Some Chinese schools and universities have reportedly implemented FRT to track attendance, manage campus facilities, and proctor exams (Tang 2021).
- Social credit system: FRT is an integral component of China's social credit system in tracking and assessing individual behavior, and in integrating various data sources, including financial records, medical histories, and travel information, to create a comprehensive system for rewarding and/or penalizing individuals (Zhou and Xiao 2019).

The Chinese use of FRT differs from that of other countries in several ways. It reflects unique approaches to the social management model, a synergistic interplay between government policies, technological innovation, and public engagement within the Chinese social context (Creemers 2017). The pervasive use of FRT underscores China's determination to leverage FRT as a comprehensive tool for enhancing national security, improving public service efficiency, and monitoring social behavior on a mass scale. This sets China apart from other democratic countries where FRT use is more limited (Kostka, Steinacker, and Meckel 2023). In addition, the Chinese government oversees the deployment of FRT and reinforces state authority over technological surveillance (Leibold 2020), government initiatives and social services (Zeng 2020). Such authority extends to data accessibility, as seen in less stringent data privacy protection when comparing China to other countries (Kostka, Steinacker, and Meckel 2021).

FRT is perceived as a population management tool in China, and has thus been considered invasive and oppressive by commentators in the West, raising concerns over privacy, human rights, and abuse of power in a climate of ubiquitous surveillance (Beraja et al. 2023; Ng 2020). It has also elicited awareness of the deep embedding of the technology into the mundane work, leisure, and organizational contexts of everyday life and the growing inability of people to opt out of biometric surveillance (Lu 2022). Some have warned of the systemic abuse of human rights through the Chinese government's deployment of FRT for monitoring dissidents and political repression (Feldstein 2019). The report of ethnic and racial discriminatory practice in Xinjiang raised serious concerns over system biases, leading to new challenges over accountability of law enforcement and the public service, and concerns for social justice (Leibold 2020; Kuttenkeuler 2021). Sufficient evidence from within China points to the fact that biases among those who design the algorithms can be introduced into the development of facial recognition technologies, leading to discrimination and inaccurate identifications of individuals (Zhang 2021; Shen 2018). In addition, the use of "emotion-recognition" in which facial traits such as happiness, boredom, and sadness are tracked to determine a person's feelings raises grave apprehensions regarding the emergence of an Orwellian dystopia (Keegan 2019; Standaert 2021). Moreover, vast amounts of data collected by FRT create a significant cybersecurity risk in the event of identity theft, data hacking or other security breaches (Yu 2023). Although these concerns are not uniquely Chinese, they are more prevalent in China due to its restrictive public sphere and the potential scale of the impact.

Within China, there are different interpretations of FRT. While the majority of the Chinese population may be indifferent to the concept of privacy, an increasing number of urban citizens have identified ethical issues such as "privacy" in various legal cases against the use of FRT (Liu et al. 2021). The extensive use of FRT in public spaces has the capacity to rewrite what it means to remain anonymous in public places. Existing research on the ethical aspects of the use of FRT for public surveillance point to the tricky balance between achieving security and protecting people's privacy and civil liberties (Brey 2004; Smith and Miller 2022). The Chinese case raises further questions regarding whether ethical problems of privacy and function creep should be ignored to achieve enhanced forms of trust and security. Coupled with other forms of bio-sensing capabilities (such as those identifying emotions, fingerprints, or body temperature), FRT raises fears among the public of a surveillance society within which the mishandling of integrated datasets by government and/or private companies could lead to serious harm to civil rights and freedom of speech (Pozen 2005; Barrett 2020; Fletcher 2023).

While the government acknowledges the concerns over data security and privacy infringement in commercial settings, it tends to ignore more serious human rights concerns in relation to the use of FRT by government departments (Nicholas 2018). FRT is just another Chinese surveillance society measure, it is argued, while others claim that recent regulatory measures can be understood, at best, as mobilizing the rhetoric of social cohesion and national security to circumvent a ban on the FRT use based on human rights violations (Hao and Lin 2023).

This chapter focuses on a residential compound in Shanghai that is attempting to implement facial recognition technology for entry control but has faced substantial opposition from its residents. Our survey with local residents focuses on the public's attitudes, approval, and concerns about FRT. Our study highlights the significance of social dynamics in influencing individual acceptance or opposition to FRT. While existing research based on public surveys in China tends to portray FRT as being influenced by convenience, efficiency, and the government's role in framing public discourse (Kostka, Steinacker, and Meckel 2020), our findings demonstrate that social dynamics such as class, gender, technological literacy, and political affiliation play an important role in shaping public approval of FRT. This derives in part from FRT's ambiguous capability; that is, the anticipated advantages of FRT in specific scenarios might be seen as disadvantages in others.

Whilst the survey captures individual attitudes, our participant observation in a WeChat group established by the residential committee offers a more nuanced understanding of how these opinions manifest in everyday interactions and decision-making processes. It provides crucial qualitative data, capturing the subtleties of how FRT influences social behavior and governance at a micro level.

This chapter also investigates key government documents, media reports, and regulatory materials which serve to contextualize these individual and communal experiences within the larger national narrative of FRT. By examining how government frames FRT in terms of convenience, security, and accountability, we can assess the state's role in shaping public perceptions and the sociotechnical imaginaries surrounding FRT. These methods create a multi-dimensional exploration of how FRT is experienced, debated, and integrated into both public and private life.

In the sections that follow, we introduce the challenges inherent in defining FRT's capabilities, especially in relation to the contrasting levels of automated decision-making and human agency determined by different systems. The ambiguity in defining levels of human control makes the constructing of "sociotechnical imaginaries" particularly important in shaping public perception of FRT.

A Precarious Equilibrium: Sociotechnical Imaginaries Through the Lens of FRT

The multifaceted nature of the technology and its diverse uses have resulted in considerable ambiguity when trying to achieve a comprehensive and universally accepted FRT definition. One of the primary challenges stems from its rapid evolution in terms of algorithmic design and technical capabilities. The application of FRT can range from geometric-based face recognition to more recent applications in deep learning, where features such as age, gender, ethnic origin, or even emotional state are claimed to be capturable, detectable, and readable (Mann and Smith 2017).

Evolving capabilities of FRT makes it difficult to define what it can or cannot do, and there is much speculation and promotional propaganda from those interested in selling or justifying the use of FRT in various contexts (Gates 2011; Rezende 2020), including in popular television or cinematic representations which tend to hype up or accentuate its operational capabilities (Singer 2011).

Defining where human influence ends and algorithmic autonomy begins is a challenging task in relation to FRT. As an adaptive technology working with critical human intervention, it is difficult to determine the mix of self-directed and human-guided FRT learning (Selwyn et al. 2023). Human interference in FRT can extend beyond the selection and processing of training datasets (Wu et al. 2022). Distinguishing between biased outcomes caused by algorithmic autonomy and those that stem from the role of human agency in the design and curation of datasets can be challenging. In areas such as national security and law enforcement, the question of human agency is increasingly vital for understanding the level of human intervention and accountability required to govern these technologies (Raposo 2022).

Ambiguity in terminology persists even when only the automated decision-making aspect of the technology is considered. Automation, in this context, often implies the ability to perform tasks or make decisions without direct human intervention. FRT's current use raises questions about the extent to which it operates independently from the parameters guiding its decision-making and its human oversight (Andrejevic and Selwyn 2020). It relies on algorithms which enable machines to compare and analyze facial features with remarkable speed and accuracy, but these processes are often directed by those who deployed the technologies in the first place (Wu et al. 2022). Human intervention is a critical part of FRT's operation from algorithmic design to making decisions on error correction.

More importantly, if FRT is to be trusted with making decisions in an autonomous manner, that is, independently, does that mean those who deploy the machine can walk free from responsibility for any errors, biases, or other discriminatory outcomes?

The lack of clarity around FRT's automated decision-making capability has exacerbated concerns about mass surveillance, privacy, and diminution of civil liberties. Some argue that only when understanding the level of human involvement in the design, deployment, and data processing capability of FRT can issues concerning regulatory liability be addressed (Mann and Smith 2017; Almeida et al. 2022; Smith and Miller 2022). As a result, deployment of FRT encounters substantial public resistance. Concerns regarding unwarranted surveillance, data breaches, and the potential misuse of personal information create a palpable sense of unease among the public, leading to heightened skepticism.

Addressing these concerns requires open and transparent dialogue between policymakers and the public in order to cultivate a sense of accountability and trust regarding the technology's intended use. The concept of "sociotechnical imaginaries" is essential here in comprehending the attempt by authorities charged with building public consensus and confidence in FRT through the deployment of ethical guidelines and regulatory mechanisms.

Sociotechnical imaginaries refer to the "collectively imagined forms of social life and social order reflected in the design and fulfillment of nation-specific scientific and/or technological projects" (Jasanoff and Kim 2009, 120). In a more recent expansion of the concept to encompass new technologies that are less bounded within nation states, Jasanoff and Kim suggest that

> multiple imaginaries can coexist within a society in tension or in a productive dialectical relationship. It often falls to legislatures, courts, the media, or other institutions of power to elevate some imagined futures above others, according them a dominant position for policy purposes. Imaginaries, moreover, encode not only visions of what is attainable through science and technology but also of how life ought, or ought not, to be lived; in this respect they express a society's shared understandings of good and evil (Jasanoff and Kim 2015: 4).

Traversing diverse societal domains, FRT-enacted sociotechnical imaginaries have embodied reflections on the potential benefits of enhanced security and convenience, as well as on issues of surveillance, privacy infringement, and other ethical considerations; as such, they necessitate navigating a precarious equilibrium between the "good" and the "evil".

What makes FRT unique in comprehending the Chinese "sociotechnical imaginaries" is how the precarious equilibrium is achieved through the wide adoption of such technologies by the government for the purpose of social control alongside the regulation of the same use in commercial settings (Chen et al. 2020). To create a balance between social control and social engagement has always been at the heart of the PRC's social governance model (Brazelton 2019). While highlighting FRT's potential for addressing persistent national security issues, a trade-off between efficiency, individual privacy, and national security is underscored. Meanwhile, the emphasis on the necessity for transparent and accountable public services accentuates the autonomous capabilities of FRT (Henman 2020).

Private companies involved in the development of the technology tend to emphasize the discourse of efficiency, security, trust, and convenience, while downplaying ethical dilemmas related to algorithmic biases and discriminatory profiling (Kitchin 2016). FRT is widely used in commercial settings, including for making payments, accessing services, or shopping online, by simply scanning faces. For individuals, FRT offers convenience and security for access control in residential complexes, offices, and public places. FRT is used for ticketing on public transportation systems and hotel check-ins, improving efficiency and reducing administrative burdens for organizations (Xu and Xiao 2018). FRT is integrated into various social media platforms for adding filters, stickers, and special effects in photos and videos. More recently, some retailers have begun to use FRT for customer identification and personalized shopping experiences which can offer product recommendations. Its uses in these commercial settings have been extended to public service provision with the aim of preventing errors, enhancing security measures, and improving convenience.

Achieving an equilibrium concerning the deployment of FRT is based ultimately on interpreting the technology's capabilities and deciding which aspects to consider when it comes to regulating those capabilities. While current regulation of FRT focuses on restricting its use in commercial settings, the power of the government in interpreting FRT's capabilities remains uncontested. It is important to note that such power in shaping the meaning of the technology has become a key mechanism of the Chinese surveillance model. As some argue, FRT is unlike traditional surveillance methods, so it is challenging for the public to fully comprehend how ethical considerations might be dealt with effectively either by the government or by private entities (Smith and Miller 2022).

The regulation of FRT is of paramount importance to the Chinese government, which recognizes its key challenges as well as its benefits in achieving the precarious equilibrium between social control and social engagement. A closer look at recent national policy, regulation, and guidelines on FRT illustrates this very issue. Under the government's responsible AI development proposal, regulation of FRT reflects an attempt to address some of the ethical concerns of the public regarding the rapid deployment of AI technologies in various areas of Chinese society. These include effective regulations to safeguard individual rights and to satisfy ethical concerns, to prevent data leaks and secure critical infrastructures, to ensure fair competitiveness and market practices, to enhance consumer protection, and to adhere to global standards of AI development (Zhu 2022; Roberts et al. 2021).

There are many regulatory documents and guidelines developed to date that address public concerns about FRT, either published by the High Court, the People's Congress, or by local government. We reviewed 10 key documents (see Table 5.1) that have been in the public domain since 2016.

Table 5.1 Overview of published Chinese documents concerning the regulation of facial recognition technologies 2016–2023

Name of document	Date of Publication	Issuing body	Key issues addressed by the document
Cyber Security Law of the People's Republic of China	Nov 7, 2016	Law: The 12th National People's Congress, the 24th Meeting	Defines "personal information": refers to various information recorded electronically or in other ways that can identify a natural person's personal identity alone or in combination with other information, including but not limited to the natural person's name, date of birth, ID number, personal biometric information, address, phone number, and so on. (Chapter 7, Item 76, No. 5)
Information Security Technology Facial Recognition Data Security Requirement	April 28, 2020	Guideline: The State Administration for Market Regulation and the National Standardization Administration	Defines the function, performance and security requirements of information systems that use FRT to perform remote identification (No. 8.1.3)
Civil Code	June 2, 2020	Law: The 13th National People's Congress, the 3rd Meeting	Same as in Cyber Security Law but adding that "information processors should take necessary measures to ensure the security of personal information they collect and store, and prevent information leakage, tampering, and loss ... Provisions are made to inform natural persons and report to the relevant authorities." (No. 1034)

Name of document	Date of Publication	Issuing body	Key issues addressed by the document
Information security technology – personal information security specification	October 1, 2020	Guideline: The State Administration for Market Regulation and the National Standardization Administration	Puts in place the ruling that "before collecting personal biometric information, informed consent should be obtained" (No 5.4) and "personal biometric information should be stored separately from personal identifiable information". In principle, the document suggested that "biometric information should not be stored and if stored, it should be of a summary and non-traceable nature and should be deleted after the authentication and identification functions are fulfilled."
Public Consultation on the Information Security Technology Facial Recognition Data Security Requirement	April 23, 2021	Guideline: National Information Security Standardization Technical Committee	In principle, FRT should not be used to identify minors under the age of 14, and FRT should not be used for purposes other than identification. FRT cannot be used to evaluate or predict work performance, financial status, health, and other preferential uses.
Data security law of the People's Republic of China	June 10, 2021	Law: The 13th National People's Congress, the 29th Meeting	The state establishes a data classification based on the importance of data in economic and social development...and the protection of data according to national security, public interests, individuals, and organizations. (No. 21)
Several Issues Concerning the Application of Law in the Handling of Criminal Cases of Violation of Citizens' Personal Information	July 27, 2021	Judicial interpretation of law: Supreme People's Court; Supreme People's Procuratorate	When residential property uses FRT as a form of access control, it shall obtain the consent of the individual owner of the property. If there is no consent, the property management company shall provide an alternative verification method. (No. 10, Item no. 1)

Name of document	Date of Publication	Issuing body	Key issues addressed by the document
Personal Information Protection Law of the People's Republic of China	August 20, 2021	Law: The 13th National People's Congress, the 30th Meeting	The installation of image acquisition and personal identification equipment in public places shall be necessary to maintain public safety and comply with relevant national regulations. It shall be set up with prominent reminder signs. The collected personal images and identification information can only be used for the purpose of maintaining public safety, and shall not be used for other purposes; unless individual consent is obtained. (Article 26)
Regulations on the Administration of Internet Data Security (Draft for comment)	November 14, 2021	Guideline: The State Internet Information Office	Data processors shall not use biometric features such as faces as the only means of personal identification or force individuals to agree to the collection of their personal biometric information (No. 25)
Application of Facial Recognition Technology Safety Management Regulation (Trial) (Draft for Comments)	August 8, 2023	Public Consultation: Cyberspace Administration of China	Article 4. Facial recognition technology may be used to process facial information only if it has a specific purpose and sufficient necessity, and when strict protective measures are taken. If there are non-biometric identification technology solutions available which can achieve the same purpose or meet equivalent business requirements, the non-biometric identification technology solution should be given priority.

As shown in Table 5.1, the acceptable and appropriate use of FRT in China must satisfy the following: (1) necessity, considering whether there are alternatives; (2) data security measures, considering the possible impact of FRT, technical measures to strengthen risk prevention, and encryption measures to ensure data security; (3) individual consent conditions, including providing clear notification, collecting the minimum necessary information, storing the original biometric data separately, only saving the summary information, and deleting the original biometric information.

Since 2020, China has adopted much tougher regulations for FRT use, aligning with international standards such as the European Union's General Data Protection Regulation (GDPR).[26] For example, the "Information Security Technology Personal Information Security Specification" included the individual consent clause as well as suggesting that "biometric information" should not, in principle, be stored. The 2023 "Facial Recognition Technology Safety Management Regulations" by the Cyberspace Administration of China further strengthens the regulation around data storage by stating that biometric data such as facial images can only be saved by personal information processors for legal purposes and that the individual's consent must be obtained separately.

The regulation of FRT in China does not necessarily reflect how the technology is used and understood at the grassroots level, as will be investigated later; rather, it illustrates China's elite-driven techno-nationalistic imaginaries of the future.

Techno-nationalistic Imaginaries of FRT and China's AI Future

In the earliest stages of the development of FRT in China, the state engaged in active identification and promotion of FRT and its capabilities, noting its ability to enhance public administration efficiency on a range of fronts. More recently, such an approach has been replaced by the regulatory question and the role of the state within it (Hao and Lin 2023). These various approaches reflect the broad sociotechnical imaginaries with reference to the terminological uncertainties surrounding FRT, its functionality, and an imaginary of its potential future uses.

Historically, the Chinese government has taken advantage of sociotechnical imaginaries to influence how people perceive and approach the integration of technology into various aspects of life. The catching up to and then surpassing of Western competition has been a key component of the sociotechnical imaginaries based on the renewed articulation of a particular world order discourse and a sense of victimization and disempowerment of China at the hands of Westernized representations (Mu and Keun 2005). In the case of FRT, this "underdog" discourse prioritizes the technology's capabilities in maintaining national security, economic development, and international domination, all of which are deeply intertwined with the country's geopolitical ambitions.

[26] GDPR (General Data Protection Regulation) is a regulation on information privacy issued by the European Union in 2016 to establish general obligations of data controllers and their processers, including the obligation to implement appropriate security measures according to the level of risk. https://gdpr-info.eu/

In terms of national security, the government AI discourse focuses on the legitimacy of the state's massive investment and leadership in technology advancement associated with FRT, portraying it as indispensable in defending border security, ensuring domestic order, and supporting national utility (Acharya and Arnold 2019). The government has been actively promoting FRT research, including it in its national strategic plans such as "Made in China 2025",[27] the new generation artificial intelligence development plan (AIDP),[28] and the AI security white paper, published annually by the China Academy of Information and Communication Technology (CAICT).[29] Investment in AI is estimated to be in the tens of billions of dollars (Acharya and Arnold 2019), while in 2017 the Shenzhen Security China Expo featured FRT and its applications as key to the upgrading of national security measures in the AI era.[30]

China's sociotechnical imaginaries about FRT underscore a particular techno-nationalism based on an evolutionary view of the country's global status from a position of perceived obscurity to a global superpower in AI research and innovation (Plantin and de Seta 2019). Within this evolution, China seeks to achieve technical sovereignty in AI development, reducing its reliance on foreign technologies (Kokas 2023). FRT is also part of the military modernization narrative which prioritizes advancements in autonomous cyber defense capabilities and intelligence-gathering to safeguard its broader national security (Allen 2019).

The deployment of FRT aligns with these national security goals (Roberts et al. 2021). The 2017 Artificial Intelligence Development Plan (AIDP)[31] illustrates a trend in the government to label AI, including FRT, as a weapon in a new round of

[27] Made in China 2025 is a 2015 national strategy to promote China's leading status as a future manufacturing superpower and AI is one of the key areas identified. https://www.gov.cn/zhuanti/2016/MadeinChina2025-plan/

[28] The State Council Notice of the New Generation Artificial Intelligence Development Plan, July 20, 2017, State Development, No. 35. http://www.gov.cn/zhengce/content/2017-07/20/content_5211996.htm For a full translation of AIDP 2017, see https://digichina.stanford.edu/work/full-translation-chinas-new-generation-artificial-intelligence-development-plan-2017/

[29] China Academy of Information and Communication Technology published an AI security whitepaper: http://www.caict.ac.cn/kxyj/qwfb/bps/201809/P020180918473525332978.pdf

[30] China International Public Security Products Expo (or 'Security China') is a government sponsored industry exhibition with the aim to enhance international collaboration and innovation in the development of a security industry. It showcases advanced technologies, leading industry practices and market trends. https://www.securitychina.com.cn/

[31] The State Council Notice of the New Generation Artificial Intelligence Development Plan, July 20, 2017, State Development, No. 35. http://www.gov.cn/zhengce/content/2017-07/20/content_5211996.htm For a full translation of AIDP 2017, see https://digichina.stanford.edu/work/full-translation-chinas-new-generation-artificial-intelligence-development-plan-2017/

technology-enabled warfare. Wang Zhigang, secretary of the partner leadership group and vice minister of the Ministry of Science and Technology, suggests that[32]

> As a strategic technology leading the future, artificial intelligence has become an important weapon for the world's major developed countries to enhance their national competitiveness and maintain national security.

China's strategic approach to state-led economic development in the realms of AI and FRT is predominantly facilitated through the implementation of its Five-Year Plans which serve as blueprints for aligning national priorities and catalyzing targeted growth in areas critical to China's global competitiveness (Ding 2018). Among the leadership group, China is believed to have early mover advantages in areas such as speech recognition and visual recognition technologies (Li, Tong and Xiao 2021). China is also believed to have surpassed many developed countries in the exportation of FRT technologies (Knight 2023).

China's endeavors in AI technological innovation, including those in FRT, are integral to its aspiration to lead global development, an aspiration not solely concerned with global competition but also with foregrounding China's rise to prominence as a world-leading economy and associated polity that purportedly operationalizes state commitments to enhanced agility in the delivery of public administration and achievement of social cohesiveness (Bilgin and Loh 2021).

This position is reflected in the CAICT's 2021 report which outlined a model for operationalizing social trust based on a range of inherent attributes of AI technologies such as security, reliability, and accountability (CAICT 2021, 6). The CAICT's report positions technological progress as a "double-edged sword"; that is, on the one hand, the technology provides new tools and approaches to enhance capacity in managing the security of national cyberspace and improving the prevention of major national risks, while at the same time those technologies could invoke a wide range of issues and vulnerabilities related to legal, ethical, social, economic, and political securities. The concept of "credibility" thus entails a balancing act between deployment of the technology as a form of social control and regulation of the same technology for the purpose of social engagement. As we shall see later, the state places greater importance on cultivating a collective view of what FRT can and cannot do by defining its technological capability, usages, and its regulations in relation to concerns of the public.

[32] Wang was interviewed by Xinhua News Agency on 21/07/2017, which was published as 'Construct the first-mover advantage of artificial intelligence, grasp the strategic initiative of the new round of scientific and technological revolution' http://www.gov.cn/zhengce/2017-07/21/content_5212404.htm

Public Discourses on FRT

The use of FRT has given rise to a high level of public concern over ethical issues such as consent, accuracy, and bias. In 2019, a law professor from Hangzhou, Guo Bing, filed a lawsuit against Hangzhou Wildlife World over the use of FRT as the only means of access to its facility, and this has become known as the "first case of facial recognition" in China (BBC 2019). This legal case informed new rounds of public debate concerning privacy as a basic human right, accompanied by a shared concern that such technologies are changing the very nature of Chinese public life (Li 2019). It prompted Chinese regulators to issue new rules to oversee the use and management of FRT.

In response to China's first FRT legal case, the Cyber Security Law was amended by stating that the collection of personal information must be "lawful, legitimate and necessary"; however, the civil court in the final ruling of this case cited Consumer Protection Law to justify the use of biometric technology by Wildlife World as ensuring accuracy and convenience. The court held that the data collected by Wildlife World exceeded the requirements of the principle of necessity and had the potential to violate personal privacy and cause harm (Lin, Du, and Liu 2021). This case was a precursor to China's special law on personal information protection and related judicial interpretations.

The case is one of a few involving FRT in citizens' everyday lives. Amid heightened security worries and increasing labor expenses, numerous residential compounds in China have turned to implementing FRT since 2015, adding to ongoing public discourse on maintaining the precarious equilibrium between security and convenience. Research by Kostka, Steinacker and Mechel (2021) found that acceptance of the use of FRT technology in public was much higher in China than in the UK, USA, or Germany (the countries covered in their survey). Significantly, their work demonstrated much greater trust in the government on the part of Chinese respondents than in the other surveyed countries.

To understand Chinese people's attitudes toward the omnipresent FRT-equipped security cameras at the time of the COVID emergency, we conducted a small survey with a residential compound in Shanghai in November 2021. A total of 76 valid survey results were considered (out of the 150 collected), indicating a response rate of approximately 51%. Although the numbers were small, the survey was supplemented by an analysis of WeChat comments in the residential group that provided insight into the range of ways in which Chinese residents approached the issues raised by the deployment of FRT technology.

Choosing a residential setting as a primary use case for the study of FRT offers an insight into the application of FRT within everyday urban living. As one of the most common use cases, FRT in residential compounds extends beyond basic security functions, encompassing a range of auxiliary applications including visitor management, amenity access control, and personalized service delivery. This use case presents an ideal setting for examining the public perception of FRT and the ethical considerations associated with its proliferation in everyday public life.

Situated in one of the most upscale real estate areas, this housing complex accommodates affluent residents, generally high-earning and extensively educated professionals. The body corporate chose HIKVISION's video intercom units with inbuilt facial recognition features as their new access control system. The system had been installed but its FRT capability had not been activated due to ongoing challenges from the residents. The main concerns among residents were about privacy and data security. In early 2021, a Wechat group was established to facilitate the exchange of opinions among residents regarding the implementation of FRT. The question of whether FRT could be activated attracted over 10,000 comments on the platform between June and December 2021.

The survey results indicate a minimal level of approval of FRT among the participants (N=76) who had exposure to the technology, whether in private or public settings, before the survey. Among the most common use cases experienced by the participants, public uses including public services, banks, and public transport received most approvals from the participants as opposed to private uses, including in residential settings, sports stadiums, and workplaces (see Table 5.2).

Table 5.2 Public acceptance of FRT based on a residential survey in Shanghai

Social acceptance of FRT	No. of respondents	Percentage
Residential	4	5.26%
Shopping Malls	19	25%
Mobile apps	13	17.11%
Sports stadiums	8	10.53%
Public Transport	36	47.37%
Workplace	14	18.42%
Public services	45	59.21%
Banks	47	61.84%
Others	3	3.95%
Total	76	

The survey confirms that public perception of FRT was multifaceted, reflecting a wide range of often conflicting perspectives and opinions as a result of the ambiguity in determining the extent of human oversight, as described earlier. While the technology offers convenience in terms of access control and authentication, there are strong concerns over 'privacy', 'ethics', and 'data security'. It was seen as a tool to encourage 'responsible conduct', but faced criticism for its potential encroachment on 'personal data security' and 'social trust'. For example:

> How can the elderly understand these new things? They can stand there for a long time without being able to read it, and they have to call for help. (68)

> On rainy days or when the camera is backlit, it often cannot be recognized, and you have to queue up to enter the community during peak hours after getting off work. (35)

> I have never seen such stupid and useless artificial garbage in my life. It is a cerebral palsy thing that is specifically designed to cause trouble for the residents. (70)

> Facial recognition security needs to be improved. It should not be used in privacy scenarios such as communities, and should only be used in scenarios such as banks, governments, and companies. (40)

> Facial recognition without supervision can easily be used by people with ulterior motives. Who has the right to ask for your face? I don't trust the real estate management to be responsible for my personal data. (57)

The public response to this use case indicates also that individual perception is highly dependent on a socially-shaped understanding of FRT, influenced in part by the government's profiling of good uses and on the regulatory system which determines risk levels. The participants seemed to understand very well the trade-offs of the technology in this particular use context, ranking convenience, labor cost reduction, and security as the top benefits of FRT, while opposing the implementation of FRT on the basis of privacy infringement, misuse and mismanagement of data by the body corporate, and the high cost of system maintenance.

The very low acceptance rate, however, throws into doubt the important finding in the Kostka, Steinacker, and Meckel (2023) survey which uncovered positive associations and high approval rating of FRT among "the younger, highly educated and higher income population" (684), arguing that, despite the more skeptical view of surveillance technologies held by this demographic group, their individualized interpretations of FRT as offering security and convenience were more impactful in shaping FRT acceptance. To Kostka, Steinacker, and Meckel (2023), this suggests the importance of public framing of FRT in China in terms of efficiency and convenience, with citizens in China tending to be more "techno-optimistic" than citizens elsewhere.

Our findings suggest that such techno-optimism may differ across different social and cultural contexts. The singular focus on trust in Kostka et al.'s (2023) survey may overlook the complex social dynamics such as class, gender, or age and cultural specificities that also shape public attitudes toward FRT. Although they did suggest that these dynamics are crucial in understanding why certain groups may be more or less accepting of FRT. For instance, wealthier or more technologically literate individuals may be more willing to embrace surveillance technologies, while others such as rural residents or older generations may feel more alienated about the implication of mass surveillance. The aspect of FRT that might have been overlooked by Kostka et al. (2023) is the resignation or passive compliance as factors behind the higher acceptance of FRT. Chinese citizens might accept FRT not out of genuine support but because they feel powerless to contest the state's use of such technologies. This form of compliance could be driven more by a sense of fatalism than by trust. Furthermore, social stratification affects opinions toward FRT in China. Elite groups such as business owners or government officials may be more supportive of FRT because they perceive it as a tool of maintaining their social status, while others may exhibit deeper concerns over manipulation or misuse by authorities. Thus, the Chinese FRT narrative cannot be understood solely through the lens of trust in the government but must be viewed in light of the broader sociopolitical environment, including the sense of resignation, social control and normalization of surveillance in everyday life.

Our survey revealed that participants perceived a significant degree of human oversight in relation to FRT use in a micro social context, expressing concerns about the potential misuse and mishandling of their personal data by the body corporate providing FRT services. As comments shared in the Wechat group further suggested, the middle-class residents of the housing complex exhibited low trust in the body corporate's providing of FRT due to concerns about the competence of its employees, specifically worries about "under-educated" employees handling sensitive data. Such class-based distrust influences how individuals perceive the security and ethical standards associated with FRT, affecting their overall approval of the technology.

Our survey results suggest that social dynamics shape public perception of FRT, affecting government regulation of the technologies as well as reflecting asymmetrical power in the development of sociotechnical imaginaries. In relation to the latter, we can observe a strong sense of unity among middle-class residents regarding ethics and privacy concerns surrounding FRT use, concerns which have drawn more attention from the Chinese regulatory framework than have issues of bias which disproportionately affect the underclass.

This local study also illustrates the contrasts between tech companies' emphasis on efficiency and convenience, government discourse on security and techno-nationalism, and Chinese urban residents' attitudes toward privacy and technological abuse of power. It suggests a contradiction between FRT regulation and implementation. As illustrated through the first FRT civil case in China, individual rights cannot be defended easily, partly because it is the Chinese state regulators that are legally liable in cases against private corporations in relation to the misuse of biometric data.

Discussion

Our research findings suggest the importance of contextualizing the response to the deployment of FRT. There is not one single use of the technology, but a wide range of uses across jurisdictions and contexts. Facial recognition for building security can be viewed and understood in different terms from the state use of FRT for national security. It is one thing to support the government use of FRT for purposes already framed by state actors as necessary and legitimate, and something altogether different to respond to the local deployment of the technology in one's residential complex. The government use can seem somewhat more distanced and abstract than local deployments reliant upon security personnel about whom residents are able to make assumptions and raise concerns. High levels of trust in the government use of surveillance would be more conducive to supporting state use of the technology, whereas concerns about local corruption or incompetence could affect local uses, such as in residential complexes. Nonetheless, concerns raised at the local level could scale up depending on how the technology is deployed. Instances of corruption or data breaches could escalate concerns from the local level to less proximate ones. With increasing deployment of FRT overriding the "not the only means" test principle stipulated in the recent regulations, misuses by law enforcement and private entities that go unchallenged raise serious issues of accountability that could affect public attitudes toward the use of the technology.

Defining FRT has been a challenge because the level of human oversight is context-dependent rather than universally defined. Different societies may perceive FRT in contrasting ways based on existing values and interpretations. Our residential compound use case survey suggests that social dynamics regarding class and human agency could affect public perception and in turn influence the approval of FRT.

It is the "human oversight" in achieving transparency and efficiency that helps the state to legitimize its use by government departments and law enforcement (Li et al. 2023), but the same "human oversight" could also become a source of distrust based on ethical considerations of privacy and data security, as our survey has shown. Addressing public concerns thus requires a delicate balance between enacting sociotechnical imaginaries based on considerations of national security, innovation, and efficiency, and implementing effective regulations to actively engage with public concerns of data security and privacy. The more recent attempts by the state to restrict the commercial use of FRT highlight the latter.

The ongoing confusion over human oversight concerning FRT persists, however. Instead of providing clarity, the state has taken advantage of a terminological openness that allows continued engagement with the public discourse while downplaying its own responsibility for misuse by government and law enforcement. The guidelines and regulations reflect public concerns around privacy and data security concerning FRT, but they are also instruments that serve political interests in further validating the state's interpretative power in shaping the sociotechnical imaginaries of FRT. There are broad claims, for example, that the use of FRT has largely enhanced transparency and efficiency of government services and that it has enhanced national security while, at the same time, privacy concerns and abuse of power continue to plague its implementation (Zhang and Dan 2018).

In March 2021, Chinese citizen Gu Cheng filed a lawsuit against the property management company of his residential compound in Tianjin, challenging the requirement that residents submit facial information for entry and exit.[33] Gu Cheng asked the property management company to delete his facial data and provide other ways for him to enter and exit the property. This case is covered by the Provisions on Several Issues Concerning the Application of Law in the Trial of Civil Cases Involving the Use of Facial Recognition Technology in the Handling of Personal Information. According to Article 10 of this legal document, if a property owner or tenant does not agree with a facial verification method, they can request that the property management company provide other reasonable verification methods; however, the court also ruled that the property company's use of FRT for access control during the COVID-19 pandemic satisfied the "necessary and legitimate" condition of use clause in the new law.

The question here is whether convenience and efficiency can be used to justify the use of FRT. This has generated conflicting interpretations in the courts. In the Guo Bing case, the court ruled that accuracy and efficiency is not sufficient to justify the use of FRT, while in the Gu Cheng case the court ruled that its use for providing efficient and convenient residential access during the COVID-19 pandemic was legal. The conflicting interpretation has led to the issuing of the "Guiding Case" document by the Supreme People's Court to provide consistency in determining whether biometric data collected by FRT belongs to the broadly defined "personal information of citizens" in Chinese criminal law.

[33] This case is known as the first facial recognition case in Tianjin. Details of the case can be found at Sohu News https://www.sohu.com/a/710424868_121687424

Based on the current jurisprudence of the Court and the government policies, it is likely that FRT use will be tightened further based on the public challenge to the above question. As shown in the Guo Bing case, the Court noted that extracting facial data without consent is an infringement to privacy as once such data is acquired, the individual's right to maintain their privacy diminishes and even disappears. However, the court ruling based on consumer law could not address the issue of the potential damage caused as to violation of human rights.

Conclusion

Widespread deployment of the FRT marks a dramatic shift in the ability to identify and track people in shared and public space, raising a host of issues regarding privacy and social control. China is of particular interest with respect to the issues raised by the technology because it has positioned itself as a global leader both in the development and state deployment of FRT. Recent advances in the development of facial recognition technology suggest that the technology will receive increasingly widespread use in the coming years.

As a form of remote, touchless identification, the technology has the advantage over other systems created by the existence and ready creation of databases of photo IDs. However, neither the deployment of the technology nor the public response to its use is monolithic or uniform. As the juxtaposition of surveys in this chapter suggests, public attitudes are context dependent, which includes the entities responsible for its use and the particular use cases. There is an ongoing process of adjustment to the technology in terms of legal regulation and public attitudes. This, we argue, justifies government intervention in maintaining an intricate balance of the degree of automation and human interaction enabled by FRT, addressing the normative question of what FRT can and cannot do.

Chinese pursuit of FRT is associated with its geopolitical interests and pursuit of technonationalism. Internationally, FRT is viewed as a symbol of technological advancement and a route to propel China as a global leader in the future. Domestically, the Chinese FRT foregrounds the balancing act of the state concerning social solidarity, economic development and national security. FRT allows the government to develop greater capacity to justify state power with its promise to protect as well as control human life. The sociotechnical imaginaries around FRT provide the Chinese government the opportunity to gather public support towards its geopolitics while enhancing its dominant position in regulating the use of FRT. Through issuing strategic policies, guidelines, and regulatory documents regarding FRT, the Chinese government acquires significant power in providing definitions and interpretations of the tech's real and imagined capabilities. In this sense, FRT reflects an extension to the traditional mass surveillance model. It legitimizes the state's role in rewriting what it means to be in public life and what it means to "reasonable expectation of privacy".

How the Chinese navigate through FRT mediated public life suggests a heightened responsibility for personal data management as well as challenges for effective regulation by the state due to their embeddedness in varied contexts. The way that people make sense of FRT is still rooted in their attempts to engage with the tech in a culturally and socially informed way. Our study of the use of the technology in one residential complex highlighted local resistance to the deployment of FRT based on concerns about the security of the data and reliability of those responsible for managing it. The residential complex case demonstrated the role played by a sense of direct control over the decision to deploy the technology by local residents. It can be contrasted to the ability of the public to respond to the deployment of FRT by higher up government authorities, where support for government priorities comes into play. Based on this, we suggest that critical interrogation of FRT through the lens of a socially and culturally responsible practice should be the priority in the deployment of FRT. This study of FRT underscores the significance of exploring public attitudes toward various use cases and contexts.

References

Acharya, Ashwin, and Zachary Arnold. 2019. "Chinese Public AI R&D Spending: Provisional Findings." *Center for Security and Emerging Technology*, Dec 2019. https://cset.georgetown.edu/wp-content/uploads/Chinese-Public-AI-RD-Spending-Provisional-Findings-1.pdf.

Almeida, Denise, Konstantin Shmarko, and Elizabeth Lomas. 2022. "The Ethics of Facial Recognition Technologies, Surveillance, and Accountability in an Age of Artificial Intelligence: A Comparative Analysis of US, EU, and UK Regulatory Frameworks." *AI and Ethics* 2 (3): 377–387.

Andrejevic, Mark, and Neil Selwyn. 2020. "Facial Recognition Technology in Schools: Critical Questions and Concerns." *Learning, Media and Technology* 45 (2): 115–128.

Barrett, Lindsey. 2020. "Ban Facial Recognition Technologies for Children-and for Everyone Else." *BUJ Sci. & Tech.* L. 26: 223–235.

BBC. 2019. "Hangzhou 'China's First Face Recognition Case': Technology, Law and International Controversy." *BBC Chinese*, November 5, 2019. https://www.bbc.com/zhongwen/simp/chinese-news-50292336.

Beraja, Martin, Andrew Kao, David Y. Yang, and Noam Yuchtman. 2023. "Exporting the Surveillance State Via Trade in AI." *Brookings*, January 12, 2023. https://www.brookings.edu/articles/exporting-the-surveillance-state-via-trade-in-ai/.

Bilgin, Efekan, and Alphonse Loh. 2021. "Techno-nationalism: China's Bid for Global Technological Leadership." *LSE*, September 28, 2021. https://blogs.lse.ac.uk/cff/2021/09/28/techno-nationalism-chinas-bid-for-global-technological-leadership/.

Bischoff, Paul. 2023. "Surveillance Camera Statistics: Which Cities Have the Most CCTV Cameras?" *Comparitech*, May 23, 2023. https://www.comparitech.com/vpn-privacy/the-worlds-most-surveilled-cities/#:~:text=China%20leads%20the%20world%20in,of%20439.07%20per%201%2C000%20people.

Brazelton, Mary Augusta. 2019. *Mass vaccination: Citizens' Bodies and State Power in Modern China.* Ithaca: Cornell University Press.

Brey, Philip. 2004. "Ethical Aspects of Facial Recognition Systems in Public Places." *Journal of Information, Communication and Ethics in Society*, 2 (2): 97–109.

Burrows, Ian. 2018. "Made in China 2025: Xi Jinping Wants to Turn China into the World's Artificial Intelligence Power." *ABC Chinese*, October 8, 2018. https://www.abc.net.au/chinese/2018-10-08/xi-jinpings-plan-to-turn-china-into-the-ai-world-leader/10351674.

Chen, Qiang, Chen Min, Wei Zhang, Ge Wang, Xiaoyue Ma, and Richard Evans. 2020. "Unpacking the Black Box: How to Promote Citizen Engagement Through Government Social Media During the COVID-19 Crisis." *Computers in Human Behavior* 110, (2020): 106380. https://doi.org/10.1016/j.chb.2020.106380.

Chen, Shixian, and Li Zheng. 2017. "What Does 'Skynet' Capture?" *People's Weekly*, no. 20, 2017. http://paper.people.com.cn/rmzk/html/2017-11/20/content_1825998.htm.

Chen, Wenhao, and Min Wang. 2023. "Regulating the Use of Facial Recognition Technology Across Borders: A Comparative Case Analysis of the European Union, the United States, and China." *Telecommunications Policy* 47 (2): 102482.

Cho, Eunsun. 2020. "The Social Credit System: Not Just Another Chinese Idiosyncrasy." *Journal of Public & International Affairs*, May 1, 2020. https://jpia.princeton.edu/news/social-credit-system-not-just-another-chinese-idiosyncrasy.

Creemers, Rogier. 2017. "Cyber China: Upgrading Propaganda, Public Opinion Work and Social Management for the Twenty-first Century." *Journal of Contemporary China*, 26 (103): 85–100.

Davis, Dave. 2021. "Ventures Inside China's 'Surveillance State'." *NPR*, January 5, 2021. https://www.npr.org/2021/01/05/953515627/facial-recognition-and-beyond-journalist-ventures-inside-chinas-surveillance-sta.

DeSmith, Christy. 2023. "Why China Has Edge on AI, What Ancient Emperors Tell Us About Xi Jinping." *The Harvard Gazette*, March 16, 2023. https://news.harvard.edu/gazette/story/2023/03/why-china-has-an-edge-on-artificial-intelligence/.

Ding, Jeffrey. 2018. *Deciphering China's AI Dream – The Context, Components, Capabilities, and Consequences of China's Strategy to Lead the World in AI.* Oxford: University of Oxford.

Fang, Jason, and Michael Walsh. 2018. "Made in China 2025: Beijing's Manufacturing Bluepring and Why the World Is Concerned." *ABC*, April 29, 2018. https://www.abc.net.au/news/2018-04-29/why-is-made-in-china-2025-making-people-angry/9702374.

Feldstein, Steven. 2019. "How Artificial Intelligence Is Reshaping Repression." *Journal of Democracy*, vol. 30 (1): 40–53.

Fletcher, Adam. 2023. "Government Surveillance and Facial Recognition in Australia: A Human Rights Analysis of Recent Developments." *Griffith Law Review* 32 (1): 30–61.

Gates, Kelly A. 2011. *Our Biometric Future: Facial Recognition Technology and the Culture of Surveillance.* New York: NYU Press.

Guan, Tianru, and Xiaotong Chen. 2023. "The Emerging Scientific Public Sphere in China's Digital Economy: Weibo Discussions on Facial Recognition Technology." *Public Understanding of Science* 32 (2): 208–223.

Hao, Karen and Liza Lin. 2023. "After Feeding Explosion of Facial Recognition, China Moves to Rein It In", *The Wall Street Journal*, Aug 8, 2023. https://www.wsj.com/articles/china-drafts-rules-for-facial-recognition-use-4953506e.

Henman, Paul. 2020. "Improving Public Services Using Artificial Intelligence: Possibilities, Pitfalls, Governance." *Asia Pacific Journal of Public Administration* 42 (4): 209–221.

Jasanoff, Sheila, and Sang-Hyun Kim. 2009. "Containing the Atom: Sociotechnical Imaginaries and Nuclear Power in the United States and South Korea." *Minerva* 47: 119–146.

Jasanoff, Sheila, and Kim Sang-hyun. 2015. *Dreamscapes of Modernity – Sociotechnical Imaginaries and the Fabrication of Power.* Chicago: The University of Chicago Press.

Keegan, Matthew. 2019. "Big Brother Is Watching: Chinese City with 2.6m Cameras Is World's Most Heavily Surveilled." *The Guardian*, December 2, 2019. https://www.theguardian.com/cities/2019/dec/02/big-brother-is-watching-chinese-city-with-26m-cameras-is-worlds-most-heavily-surveilled.

Kitchin, Rob. 2016. *Getting Smarter About Smart Cities: Improving Data Privacy and Data Security.* Dublin: Data Protection Unit, Department of the Taoiseach. https://mural.maynoothuniversity.ie/7242/1/Smart.

Knight, Will. 2023. "China Is the World's Biggest Face Recognition Dealer." *Wired*, January 24, 2023. https://www.wired.com/story/china-is-the-worlds-biggest-face-recognition-dealer/.

Kokas, Aynne. 2023. *Trafficking Data: How China Is Winning the Battle for Digital Sovereignty.* Oxford: Oxford University Press.

Kostka, Genia, Léa Steinacker, and Miriam Meckel. 2021. "Between Security and Convenience: Facial Recognition Technology in the Eyes of Citizens in China, Germany, the United Kingdom, and the United States." *Public Understanding of Science* 30 (6): 671–690.

Kostka, Genia, Léa Steinacker, and Miriam Meckel. 2023. "Under Big Brother's Watchful Eye: Cross-Country Attitudes Toward Facial Recognition Technology." *Government Information Quarterly* 40 (1): 101761.

Kuttenkeuler, Michel. 2021. "The Use of Big Data and Surveillance Technology in the Context of China's Repression of the Uyghur-minority in the Xinjiang Region." *IFAIR*, February 17, 2021. https://ifair.eu/2021/02/17/the-use-of-big-data-and-surveillance-technology-in-the-context-of-chinas-repression-of-the-uyghur-minority-in-the-xinjiang-region/.

Leibold, James. 2020. "Surveillance in China's Xinjiang Region: Ethnic Sorting, Coercion, And Inducement." *Journal of Contemporary China* 29 (121): 46–60.

Li, Daitian, Tony W. Tong, and Yanggao Xiao. 2021. "Is China Emerging as the Global Leader in AI." *Harvard Business Review*, February 18, 2021. https://hbr.org/2021/02/is-china-emerging-as-the-global-leader-in-ai.

Li, Zhizhao, Yuqing Guo, Masaru Yarime, and Xun Wu. 2023. "Policy Designs For Adaptive Governance of Disruptive Technologies: The Case of Facial Recognition Technology (FRT) in China." *Policy Design and Practice* 6 (1): 27–40.

Li, Zongxian. 2019. "Facial Recognition and Privacy: Controversial Topics of Intense Discussion in China, Hong Kong and Taiwan." *BBC Chinese*, November 7, 2019. https://www.bbc.com/zhongwen/simp/chinese-news-50328377.

Lin, Haibin, Guodong Du, and Liu Qiang. 2021. "China's First Facial Recognition Case." *China Justice Observer*, May 2, 2021. https://www.chinajusticeobserver.com/a/china-s-first-facial-recognition-case.

Liu, Yu-li, Wenjia Yan, and Bo Hu. 2021. "Resistance to Facial Recognition Payment in China: The Influence of Privacy-Related Factors." *Telecommunications Policy* 45 (5): 102155.

Lu, Alex Jiahong. 2022. "Toward Everyday Negotiation and Resistance under Data-Driven Surveillance." *Interactions* 29 (2): 34–38.

Luo, Xiang. 2023. "On the Limitation and Application of Facial Recognition Criminal Law." *Peking University Legal Information Network*, May 12, 2023. https://www.163.com/dy/article/I4IAAH2H-0530W1MT.html.

Mann, Monique, and Marcus Smith. 2017. "Automated Facial Recognition Technology: Recent Developments and Approaches to Oversight." *The University of New South Wales Law Journal*, 40 (1): 121–145. https://search.informit.org/doi/10.3316/ielapa.771179858194317.

Mu, Qing, and Keun Lee. 2005. "Knowledge Diffusion, Market Segmentation and Technological Catch-Up: The Case of the Telecommunication Industry in China." *Research Policy* 34 (6): 759–783.

Ng, Alfred. 2020. "How China Uses Facial Recognition to Control Human Behavior." *CNET*, August 11. https://www.cnet.com/news/politics/in-china-facial-recognition-public-shaming-and-control-go-hand-in-hand/.

Peters, Michael A. 2019. *The Chinese Dream: Xi Jinping Thought on Socialism with Chinese Characteristics for a New Era*. London: Routledge.

Plantin, Jean-Christophe, and Gabriele De Seta. 2019. "WeChat As Infrastructure: The Techno-Nationalist Shaping of Chinese Digital Platforms." *Chinese Journal of Communication* 12 (3): 257–273.

Pozen, David. 2005. "The Mosaic Theory, National Security, and the Freedom of Information Act." *The Yale Law Journal*, 628–679.

Qiang, Xiao. 2019. "The Road to Digital Unfreedom: President Xi's Surveillance State." *Journal of Democracy*, vol. 30 (1): 53–68.

Raposo, Vera Lúcia. 2022. "The Use of Facial Recognition Technology by Law Enforcement in Europe: A Non-Orwellian Draft Proposal." *European Journal on Criminal Policy and Research*: 1–19.

Rezende, Isadora Neroni. 2020. "Facial Recognition in Police Hands: Assessing the 'Clearview Case'from a European Perspective." *New Journal of European Criminal Law* 11 (3): 375–389.

Ripley, Nicholas. 2018. "Human Rights Crisis: Mass Internment of Muslim Ethnic Minorities in China." *Human Rights Law Commons*, Fall 2018. https://digitalcommons.wcl.american.edu/hrbregionalcoverage-fall2018/4/.

Roberts, Huw, Josh Cowls, Jessica Morley, Mariarosaria Taddeo, Vincent Wang, and Luciano Floridi. 2021. "The Chinese Approach to Artificial Intelligence: An Analysis of Policy, Ethics, and Regulation." *AI & Society* 36: 59–77.

Schmitz, Rob. 2018. "Facial Recognition in China Is Big Business As Local Governments Boost Surveillance." *NPR*. April 3, 2018. https://www.npr.org/sections/parallels/2018/04/03/598012923/facial-recognition-in-china-is-big-business-as-local-governments-boost-surveilla.

Selwyn, Neil, Chris O'Neill, Gavin Smith, Mark Andrejevic, and Xin Gu. 2023. "A Necessary Evil? The Rise of Online Exam Proctoring in Australian Universities." *Media International Australia* 18 (1) (2023): 149–164.

Shen, Xinmei. 2018. "Facial Recognition Camera Catches Top Businesswoman 'Jaywalking' Because Her Face Was on a Bus." *South China Morning Post*, Novenber 22, 2018. https://www.scmp.com/abacus/culture/article/3028995/facial-recognition-camera-catches-top-businesswoman-jaywalking.

Shi, Jiayou. 2022. "International Experience and Chinese Model of Facial Recognition Governance." *People's Tribune*. February, 2022. http://paper.people.com.cn/rmlt/html/2022-02/20/content_25911476.htm.

Singer, Natasha. 2011. "Face Recognition Makes the Leap from Sci-fi." *The New York Times*, November 12, 2011. https://www.nytimes.com/2011/11/13/business/face-recognition-moves-from-sci-fi-to-social-media.html.

Smith, Marcus, and Seumas Miller. 2022. "The Ethical Application of Biometric Facial Recognition Technology." *Ai & Society*: 1–9.

Standaert, Michael. 2021. "Smile for the Camera: The Dark Side of China's Emotion-Recognition Tech." *The Guardian*, March 2, 2021. https://www.theguardian.com/global-development/2021/mar/03/china-positive-energy-emotion-surveillance-recognition-tech.

Steinacker, Léa, Miriam Meckel, Genia Kostka, and Damian Borth. 2020. "Facial Recognition: A Cross-National Survey on Public Acceptance, Privacy, and Discrimination." arXiv:2008.07275.

Sudworth, John. 2017. "How Powerful Is China's Facial Recognition Surveillance System." *BBC*, December 11, 2017. https://www.bbc.com/zhongwen/simp/chinese-news-42307561.

Tang, Yahua. 2021. "Wristbands, Headbands, and Face Recognition All Come into Play. Is It Reasonable for Schools to Monitor Students?" *Sohu.com*, January 29, 2021. https://www.sohu.com/a/447401123_120829667.

Van Noorden, Richard. 2020. "The Ethical Questions That Haunt Facial-Recognition Research." *Nature* 587 (7834): 354–359.

Walton, Greg. 2001. *China's Golden Shield: Corporations and the Development of Surveillance Technology in the People's Republic of China*. International Centre for Human Rights and Democratic Development. https://ora.ox.ac.uk/objects/uuid:084840ac-b192-407b-ab6c-f8f810310369.

Wu, Xingjiao, Luwei Xiao, Yixuan Sun, Junhang Zhang, Tianlong Ma, and Liang He. 2022. "A Survey of Human-in-the-loop for Machine Learning." *Future Generation Computer Systems* 135: 364–381.

Xu, Xiuzhong, and Bang Xiao. 2018. "Chinese Authorities Use Facial Recognition, Public Shaming to Crack Down on Jaywalking, Criminals." *ABC*, March 20, 2018. https://www.abc.net.au/news/2018-03-20/china-deploys-ai-cameras-to-tackle-jaywalkers-in-shenzhen/9567430.

Ye, Josh. 2023. "China Drafts Rules for Using Facial Recognition Technology." *Reuters*, August 8, 2023. https://www.reuters.com/technology/china-drafts-rules-using-facial-recognition-technology-2023-08-08/.

Yu, Eileen. 2023. "China Closes Record Number of Personal Data Breaches, Moots Facial Recognition Law." *ZDNET*, August 11, 2023. https://www.zdnet.com/article/china-closes-record-number-of-personal-data-breaches-moots-facial-recognition-law/.

Zeng, Jinghan. 2020. "Artificial Intelligence and China's Authoritarian Governance." *International Affairs* 96 (6): 1441–1459.

Zhang, Phoebe. 2021. "The 'CEO' Is a Man: How Chinese Artificial Intelligence Perpetuates Gender Biases." *South China Morning Post*, September 30. https://www.scmp.com/news/people-culture/social-welfare/article/3150600/ceo-man-how-chinese-artificial-intelligence.

Zhang, Yixuan, and Dan Shang. 2018. "Facial Recognition Technology: Good for Productivity." *People's Daily*, August 15, 2018. http://politics.people.com.cn/n1/2018/0815/c1001-30229330.html.

Zheng, Sarah, and Jane Zhang. 2023. "China Wants To Regulate Its Artificial Intelligence Sector Without Crushing It." *Time Magazine*, August 14, 2023. https://time.com/6304831/china-ai-regulations/.

Zhou, Christina, and Bang Xiao. 2019. "China to Expand Controversial Social Credit System to 33 Million Companies Ahead of 2020." *ABC*, September 20, 2019.

https://www.abc.net.au/news/2019-09-20/china-to-expand-controversial-social-credit-system-to-companies/11527632.

Zhu, Junhua. 2022. "AI Ethics with Chinese characteristics? Concerns and Preferred Solutions in Chinese Academia." *AI & Society*: 1–14.

CHAPTER 6

The Social Credit System as a Law-Enforcing Tool: Pillars of Local Implementation

Haemin Jee

Abstract

This chapter investigates how the social credit system is implemented by local governments and uncovers new insights regarding its uses. First, local social credit systems mainly target and punish firms rather than individual citizens. Second, these targeted firms are punished under the social credit system for violating existing laws and regulations, behaviors that were already deemed unlawful and subject to punishment by existing legal and regulatory frameworks. These core findings reveal a surprising function of the social credit system—it can be used to reinforce institutions and garner more compliance with *existing* laws and regulations from firms, rather than constricting and monitoring individual-level behavior.

Keywords: Social credit system; Local governance; Law and regulation; China

Introduction

China's social credit system has been the subject of intense scrutiny by international media and observers of Chinese politics. The idea that every citizen in China—almost 1.4 billion people—could be assigned a single score that would determine their access to government services, education, and employment has alarmed and intrigued observers around the world. Commentary has, to varying degrees, portrayed the social credit system as a technology-facilitated surveillance system designed to control individual behavior. Some journalists have called the social credit system an "Orwellian nightmare" or an example of digital authoritarianism that indulges the Chinese Communist Party's most totalitarian impulses (Kim 2017; Mozur 2019). These descriptions assert that the social credit system is intrinsically linked to China's extensive and technologically-advanced surveillance systems that will ultimately be used to facilitate targeted repression of citizens. One article states that "It's [the social credit system] probably the largest social engineering project ever attempted, a way to control and coerce more than a billion people . . . If successful, it will be the world's first digital dictatorship" (Carney 2018).

More nuanced news articles have cautioned against such extreme portrayals of the social credit system, but still situate it within a greater surveillance project of the Chinese state that includes surveillance cameras, facial recognition technology, and online censorship. These reports raise concerns that citizens with low scores are banned from flights, trains, hotels, social benefits, government jobs, and good schools. In particular, critics have raised concerns about the social credit system's repressive nature, claiming that it has been used to monitor and punish political activists and human rights lawyers. Overall, Western media sources have portrayed the social credit system as an extreme invasion of privacy and a tool for social control. Coverage of the social credit system has been dominated by stories of students with admissions to top universities being revoked due to their parents' low social credit scores, of human rights activists on no-fly blacklists, and of ordinary citizens punished for excessively playing online games. This analysis of the social credit system as a repressive surveillance tool is also echoed in some of the scholarly works on the social credit system (Hoffman 2017; Xu et al. 2020; Kostka and Antoine 2020).

There are surprisingly few clear descriptions of what the social credit system actually is, or how it operates. Is it a new technology, created through supercomputers and complex artificial intelligence algorithms? Is it an extension of online surveillance tools? Is it run by private companies, government officials, or both? To this end, it is important to identify where the social credit system has been implemented, who the key operating actors are, and the tools that constitute the pillars of implementation. As I will show, while the stated goal is national implementation, the social credit system has yet to be implemented everywhere and in its current form is geographically limited. The most important actors are local governments and their agents. A large body of literature in Chinese politics emphasizes how centrally-directed policies are variously interpreted and implemented by local officials, adding another layer of complexity to how policies are actually put into practice (Oi 1999, 2011; Montinola, Qian, and Weingast 1995; Takeuchi 2014; Tsai 2007; Chung 1995). Observers have noted that in the Xi Jinping era, anti-corruption and centralization campaigns have consolidated power in central institutions (Economy 2018; Shirk 2018), while others have challenged the idea that local experimentation under Xi is in decline (Ang 2023; Heffer and Schubert 2023). To truly understand the social credit system—its purpose, reach, and impacts—we must first examine in depth how local governments implement the social credit system. It is these local efforts that reflect the most active and relevant manifestation of the social credit system.

To this end, this chapter investigates how lower-level governments implement local social credit systems. Relying upon close readings of government documents and materials across pilot cities, this chapter describes the major institutions and programs sitting under the umbrella of the social credit system. Using more systematic data on implementation efforts, with more granular data on specific implementation efforts, I establish two major findings. First, local social credit systems target firms rather than individuals. Second, firms are targeted for violations of existing laws and regulations. These insights reveal a previously under-studied function of the social credit system—that it is used to reinforce and strengthen existing efforts to ensure compliance and to encourage lawful behavior among economic actors.

The chapter proceeds in the following manner. I first highlight how the social credit system is implemented at the local, not national, level and discuss the data collection process for examining local implementation efforts. Using data from publicly-available government documents, reports, and local news stories, I then identify the major ways in which the social credit system manifests at local levels and the specific institutions that are created under the social credit system. Next, drawing upon actual records of punishments under the social credit system, the chapter addresses who is targeted under the social credit system and which behaviors are punished. Finally, I address concerns surrounding the social credit system's potential for surveillance and repression, while highlighting its current functions as a law-enforcing mechanism.

Local Implementation of the Social Credit System: Notes on Data Collection

Following the State Council's release of a planning document in 2014, many believed the social credit system to be a unified, national system that would eventually assign all individuals in China one score, a score that would determine access to government services, travel, or education and employment opportunities. Indeed, many academic studies, while distinguishing between national and local programs, focus their empirical analysis on national data-sharing platforms (Engelmann et al. 2019; Liang et al. 2018).

In reality, the social credit system is implemented by local governments, and there is no national social credit system that assigns all citizens a single score. Contrary to reports that the social credit system is a nationwide endeavor, the main political authorities overseeing the implementation of the social credit system are local government officials. The social credit system as a policy does fall under the purview of the National Development and Reform Committee (NDRC), a powerful central-level Party committee, but, as in many instances of policymaking in China, while the NDRC sets the broad goals and ambitions, it is local officials who are tasked with the social credit system's implementation. Accordingly, while the State Council's 2014 planning outline clarified the major goals of the social credit system and provided the bare outline of how such a system should be implemented at local level, it encouraged local governments to take the initiative and create *local* social credit systems.

Because there is no national implementation of the social credit system, and because the most active manifestations of the social credit system are apparent at local levels, this chapter will focus on the local implementations of the system to examine its political and economic purpose. For this endeavor, I collected three different types of data on local implementation. First, I collected detailed data on the characteristics of the social credit system as implemented at local levels. When local governments implement the social credit system, what programs are actually created? What infrastructure and institutions are built as part of the social credit system? This is the first set of questions that will be answered through a thorough investigation of local implementation practices. Second, I collected data on the differences among local governments in their implementation of the social credit system. In particular, I highlighted the variations in who is targeted by local social credit systems. Finally, I collected more granular data using records of punishment under the social credit system from a select number of cities to further highlight which entities and behaviors are subject to regulation by the social credit system.

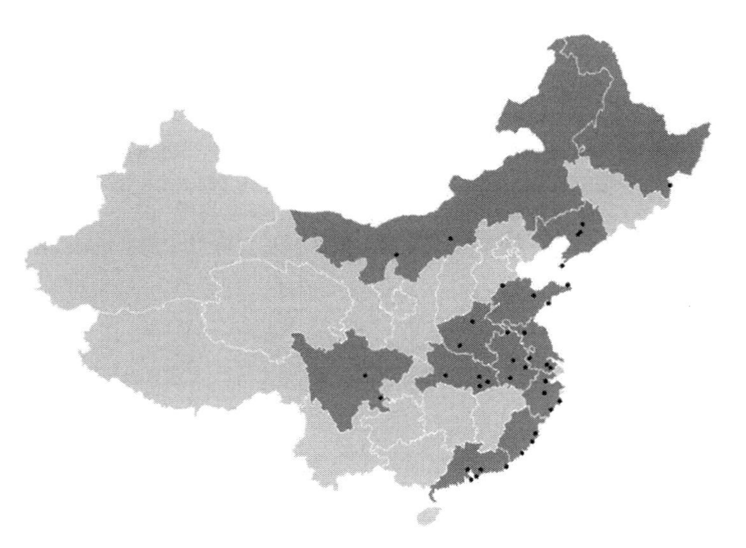

Figure 6.1 Geographic distribution of pilot cities
Note: Provinces with pilot cities are shaded in dark gray.

The total set of observations encompasses the 43 pilot cities chosen by the central government to implement the social credit system.[34] These cities represent 12 provinces from various regions of China. Figure 6.1 displays the locations of these pilot

[34] There was a new wave of pilot cities released in 2021, bringing the total to 62.

cities, with the provinces where pilot cities are located shaded in gray. Of the 43 designated pilot cities, three are district-level units in Beijing and Shanghai, three are county-level cities, and the remainder are prefecture-level cities. In addition, 11 pilot cities are provincial capitals.

Whether a city is designated as a pilot city or not is obviously not random. There are notable differences between pilot and non-pilot cities. Compared to non-pilot cities, pilot cities have higher GDP and bigger populations. Figure 6.2 demonstrates these differences in a box plot. I used logged GDP and logged population numbers from 2010, before pilot city status was determined. As demonstrated in the box plot, pilot cities have higher GDP and larger populations. Using a Welch two-sample difference in means t-test, I found that these differences are statistically significant at an level. This demonstrates that while pilot cities are located across China, they are in general more economically developed, larger, and strategically important cities. These pilot cities may have been chosen because they have the financial and human resources to implement a new program. In addition, they are considered more important or critical to regional development; central-level officials may therefore have chosen cases where the social credit system is more likely to be successful and where it would make the most impact.

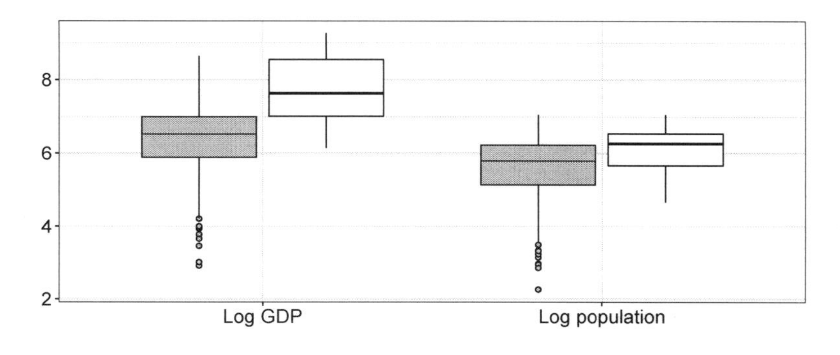

Figure 6.2 Log GDP and log population by pilot status

To theorize about the social credit system's consequences, we must first be able to describe what it is and how it functions. It is imperative, then, to identify the major programs and institutions that make up local social credit "systems."

Data Source

To investigate how local governments establish social credit systems in their own jurisdictions, I collected publicly-available documents and plans regarding local implementation. Specifically, each pilot city has a designated social credit system office, known as the SCS leadership office (社会信用体系建设工作领导小组办公室). This office sits under the city's development and reform committee, the local office of the central-level National Development and Reform Committee. The social credit system office is responsible for distributing documents pertinent to the city's social credit system, gathering and aggregating information from other city-level bureaus and offices, and maintaining the local social credit system's website. Through this website, the local social credit system office distributes key information about local implementation, uploads aggregated data on local implementation efforts, and publicizes blacklists and red lists—crucial components of the social credit system that will be discussed in later sections. Documents, announcements, local news articles, photos, and government reports were gathered from the websites of the 43 pilot city government websites.[35] These primary sources were then read and coded carefully to identify the major institutions and programs created under the umbrella of the local social credit system.

Pillars of Local Implementation

Though details of local implementation vary between localities, the primary sources examined from the 43 pilot cities reveal that there are similarities shared across local implementation efforts. I deem these to be the main *pillars of implementation*. These pillars include lists of rule violators and lists of those that display "model" behavior, ratings of firms, and individual social credit scores. The pillars of implementation evaluate the behavior of firms, organizations, and sometimes individuals. They are connected to a system of rewards and punishments that aim to shape the behavior of the relevant subjects.

Pillar 1: Lists

A key component of local social credit systems is their consolidated lists of firms, organizations, and sometimes individuals. There are two types of lists associated with local social credit systems. First, red lists are names of exemplary or "model" entities that have demonstrated particularly trustworthy behavior. Second, a

[35] More than 300 documents were gathered during this process.

blacklist is a list of rule violators. Red lists and blacklists are two sides of the same coin: red lists aim to provide incentives or benefits for good behavior (including public recognition), while blacklists aim to deter bad behavior through punishments (including public shaming). In certain pilot cities, an entry into a red list can lead to benefits, such as priority in receiving a basic living allowance or social assistance, a preferential recommendation in Party recruitment, or priority consideration when bidding for government projects. Blacklists are associated with punishments, including additional fines, bans from bidding for government contracts, and closing down of businesses, to name a few.

While both lists are interesting aspects of local social credit systems, in this chapter I will focus on the role of blacklists for a number of reasons. First, not all pilot cities have active or updated red lists, but all pilot cities have some form of active blacklist. Moreover, blacklists are the main avenue through which the social credit system is implemented. Second, the types of benefits under red lists vary widely, making it difficult to compare the meaning of red lists across localities. Finally, punishment or sanctions for rule violations may be more important in shaping incentives for compliance than are rewards (Fearon and Laitin 1996). It should also be noted that discussions of the social credit system in current academic or policy discourse focus on the punishments associated with the social credit system rather than its rewards.

Blacklists (or lists in general) under the social credit system are not an entirely new concept.[36] City-level government bureaus also maintain records of rule violators in their own jurisdictions. The office of social credit consolidates information from various government agencies to create a unified blacklist that is under its control; it is the main avenue through which the local social credit system aggregates and disseminates data on rule violators. Information about blacklist entries under the social credit system is gathered by various city-level government bureaus, including the local tax bureau, traffic police, environmental protection bureau, city planning

[36] The Supreme People's Court also maintains blacklists (失信被执行人), directly translated as a list of those who engaged in "non-performance of legally binding judgments." In 2013, the Supreme People's Court created a comprehensive blacklist system, which prevented individuals who had been obliged and capable of carrying out a valid legal order (such as a court order or administrative decision) but failed to do so from flying or buying luxury goods. Other administrative bodies at the national level began creating their own blacklists to bar individuals who had engaged in illegal behavior from train and air travel. Similar to criminal records in the United States, these blacklists were an attempt to address non-compliance with court judgments. While previous reporting of the social credit system has pointed to instances of travel bans as evidence for the repressive nature of the social credit system (Gan 2019), the travel bans on which we have information are associated with the Supreme People's Court's blacklist, not with local social credit systems. The Supreme People's Court's blacklist is not associated with that broader social credit system, even though the term "blacklist" is used in both cases.

bureau, or market regulation bureau, to name a few. The social credit system office then aggregates these various government agencies' records of rule violators to create centralized lists of rule violators. In other words, the social credit system blacklists are created from entries that originate from other government agencies' own blacklists or records of relevant rule violators. The local tax bureau, for example, submits the names of the rule-violator and the law that was violated to the local social credit system's blacklist from the information it has about tax violators. This information can then be accessed by the local bank, the labor bureau, and others. It is important to note that the social credit system blacklist entries are not new rule violators in the sense that there are no new regulations, restrictions, or rules that the social credit system is enforcing; instead, the blacklists are a compilation of known violators that respective government agencies have already collected.

Blacklists are open to the public but are mainly used to share information across government bureaus. Different government agencies can access this public social credit system blacklist and identify serial rule violators or those that have been flagged as rule violators by other government agencies. Indeed, this function of the social credit system was highlighted by a researcher involved in designing the social credit system. This interviewee stated that "The blacklists are accessed internally by the government because different agencies need information from other agencies".[37] In providing a centralized list of rule violators open to all government agencies, the social credit system makes possible information-sharing and consolidation across the respective government agencies that contribute to that list.

Pillar 2: Firm Ratings

Other than blacklists, another key pillar of the social credit system is ratings of firms. In some cities, firms or organizations are given social credit scores. These can be numerical scores (that is, a score ranging from 0 to 1000) or tiers (that is, firms placed into AAA social credit, AA social credit, B, C, and so on). Blacklists clearly identify rule-violators, but scores can be more ambiguous. Some pilot cities only have blacklists and no rating systems. In cities that have both a blacklist and a ratings system, firms with low scores are also placed on blacklists.

Pillar 3: Individual Social Credit Scores

In a select number of pilot cities, there also exists a scoring or rating system for individual citizens. These scoring systems can be points-based (that is, there is a maximum of 1000 points and points can be deducted or added) or tier-based (that is,

[37] Author's Interview BJ1801

citizens are divided into different tiers, such as AAA, AA, A, B, C, and so on). Rather than creating these scores through algorithms and big data, most citizens start with a base score or are allocated to a tier; for example, Yiwu City gives each citizen 100 points.[38] A citizen's score will then decrease if there is a record of the citizen's rule-breaking behavior. This includes behaviors such as failing to pay utility bills, traffic violations, tax evasion, or engaging in fraudulent activity. Conversely, points can be added if citizens engage in pro-social behavior, such as donating blood.

A major distinction overlooked in contemporary discussions of the social credit system is that between private credit score systems and government-run social credit systems. A common conflation is made between government social credit programs and Alibaba's Sesame Credit feature.[39] The Sesame Credit Score used on Alibaba's e-commerce platforms is calculated by Ant Financial, an affiliate of Alibaba, through an algorithm that considers an individual's purchases made through Alipay (an e-payment platform) and on Taobao or T-Mall, online shopping platforms. Rather than acting as a punitive institution, Sesame Credit uses scores to reward its users. Those with high Sesame Credit scores can, for example, borrow a ride-share bicycle without paying a deposit.[40]

Accusations that the social credit system tracks individual-level behavior in real time point to Alibaba's Sesame Credit as evidence. While Sesame Credit scores do track individual purchases on affiliated Alibaba platforms, this is similar to the way in which other technology companies (both in China and other countries) use private data to make recommendations or target ads. Crucially, an individual's commercial behavior on Alibaba platforms is not integrated with local social credit systems. Early in the development of a national social credit system, in 2015, private companies such as Alibaba were tasked with building trial social credit systems by the People's Bank of China. By 2017, however, the authority to oversee the development of a social credit system was given to the National Development and Reform Committee (NDRC). The NDRC, citing conflict of interest and lack of cooperation among participating private firms, pulled support for these private firm projects and no licenses were granted to private companies to continue building a national social credit system, though there have been some isolated reports of private companies working with local governments in their implementation of

[38] 义乌市个人信用管理 法(试行). August 10, 2017. 信用中国 (浙江·金华).

[39] Alibaba has cooperated with state institutions before; the Supreme People's Court worked with Alibaba to punish those who failed to repay debts by lowering their credit ratings on Alibaba and preventing them from buying luxury goods online. See Liu 2018.

[40] A familiar practice may be rewards programs that give consumers special benefits or ratings systems on ride-share platforms.

the social credit system.[41] Private tech companies continue to be excluded from the development of the official social credit system.[42] The Covid-19 pandemic revealed, however, that local governments have developed increased capacity to track individual movements and behaviors (need citations here about the health code). If local governments' use of digital surveillance becomes integrated with the social credit system, this could empower local governments to use social credit scores to police or regulate a wider swath of behavior, including speech on social media platforms, consumer choices, or even daily life habits, such as playing online games. It remains to be seen whether this integration between governments' use of digital surveillance and the social credit system will happen. In current manifestations, local social credit systems do not rate individuals using personal information gathered from digital surveillance.

As the above sections demonstrate, there is little evidence of the social credit system acting as an automated system. Indeed, local implementation of the social credit system reveals that the social credit system requires the time and labor of government employees to curate blacklists and ratings. Information about firms and (sometimes) individuals is aggregated and summarized, but this is not done as an automated process.

The Who and What of the Pillars

What do the main pillars of implementation reveal about the social credit system's functions? Specifically, who are the targets of the social credit system? One common assumption is that the social credit system is used to track individual-level behavior, but a close examination of local implementation practices reveals that firms, rather than individuals, are the main targets of the social credit system.

Who is targeted?

This finding is based on two major sources of data—coding of local implementation efforts across all 43 pilot cities and more granular data based on select pilot cities' blacklists. First, I coded the local social credit systems of 43 pilot cities along a few

[41] For example, Wuhu city collaborated with telecommunications companies to create a "Dishonesty" ringtone for those who had defied court orders. Whenever a call was made from or to this person's phone, the person who was making or receiving the call would hear this particular ringtone.

[42] "China's Social Credit System in 2021: From fragmentation towards integration." Mercator Institute for China Studies, March 2021.

important dimensions, using the primary sources gathered from publicly-available government websites. I coded whether a pilot city: (a) publicizes a blacklist of firms; (b) publicizes a blacklist of individuals; (c) has announced plans or a planning outline to create a personal social credit system for individuals; (d) has created a personal social credit score system for individuals; and (e) for a pilot city with a personal social credit score system, whether they had also created a smartphone application, a web portal, or a WeChat public account where individuals could access their social credit score. Figure 6.3 displays the variations among pilot cities in who is targeted by the local social credit system. Prefectures in white are non-pilot cities.

In Figure 6.3, a 1 indicates a pilot city that has implemented only a blacklist of firms. A 2 indicates a pilot city with a blacklist of firms and individuals. A 3 indicates that the pilot city has the aforementioned blacklists and has announced plans to create a personal social credit scoring system. A 4 refers to a pilot city that has already created a personal social credit scoring system as well as firm blacklists. Finally, a 5 refers to a pilot city that has implemented firm blacklists and a personal social credit scoring system, and has developed a web application, phone application, or a WeChat public account that allows individuals to access this score. I find that among prefecture-level pilot cities,[43] all have created a publicized blacklist of firms; however, out of the 37 prefecture-level pilot cities, only 14 (38%) have created a personal social credit scoring system in which residents are given a score or a rating of "trustworthiness." Among the 23 pilot cities that have not yet implemented a personal social credit scoring system, 13 have announced plans to do so. Finally, among the 14 cities that already have a personal social credit scoring system, 10 have created a web application, smartphone application, or a public account on WeChat that allows individuals to access their own social credit scores. This indicates that only 10 of the 37 prefecture-level pilot cities, or about 27%, have created an individually-targeted, personal social credit scoring system with an online component that allows individuals to access their social credit score. These types of scoring systems that give individual citizens actual scores or ratings have dominated the news coverage of the social credit system, but it remains the case that most of the local implementation efforts have focused primarily on regulating the behavior of firms.

[43] Administrative jurisdictions in China are province, prefecture, and county, in order of decreasing size.

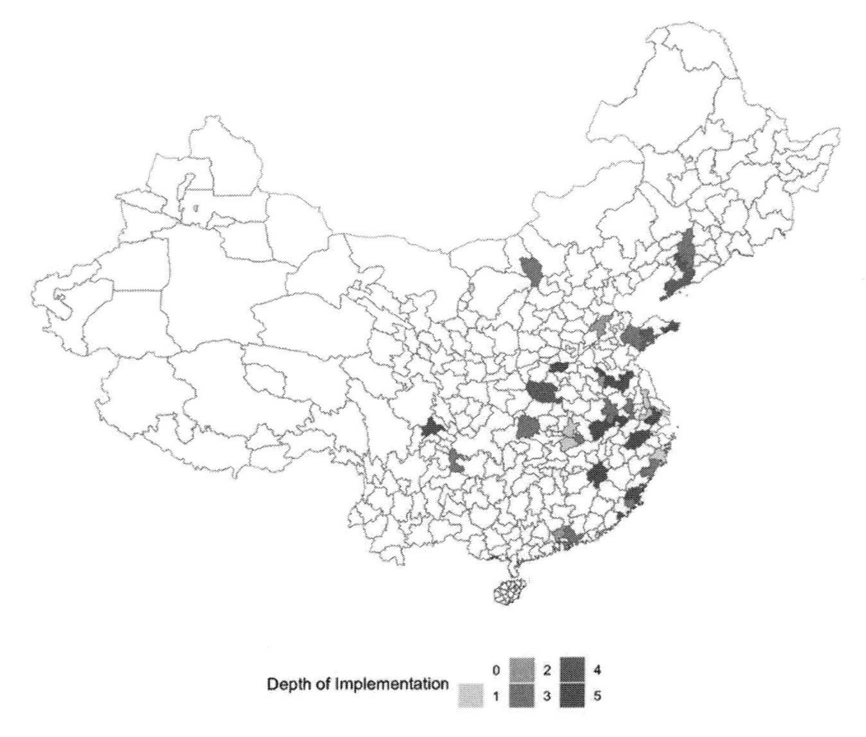

Figure 6.3 Varying Levels of Local Implementation

The finding that local social credit systems target firms rather than individuals is also supported by evidence from local blacklists. I collected blacklist entries from four pilot cities that represent major regions of China (North, South, West, and East). While Eastern China is over-represented among the pilot cities, I deliberately chose to collect blacklists from pilot cities in other regions to provide a more representative picture of local social credit system implementation. Further, because a city's GDP per capita is strongly correlated with government expenditure and local state capacity, I chose cities that differed in their level of economic development. The four cities are Hangzhou, Fuzhou, Zhuhai, and Hefei.

Using web scraping tools, I collected all the blacklist entries of these four selected cities. Each blacklist entry includes: the name of the entity placed on the blacklist; when the blacklist entry was created; the reason for the entry; and the government bureau that uploaded or submitted the blacklist entry. Using this data, I was able to extract the number of blacklist entries across the four cities that named individuals.

Table 6.1 Individuals on Blacklists

City	% Entries that are individuals
Hangzhou	31%
Zhuhai	0.1%
Fuzhou	<0.1%
Hefei	None

Table 6.1 displays the percentage of entries on four local blacklists that are individuals. Three of the four pilot cities do include individuals on blacklists, ranging from 31% to less than 0.1% of blacklist entries.[44] One pilot city includes no individuals on the blacklist. This is consistent with other investigations of blacklists (Brussee 2022). Overall, both the data indicating local variations across all pilot cities and the data showing certain pilot city implementation efforts reveal that firms are the main targets of social credit systems, not individuals.

This finding raises doubts about the social credit system being used primarily to track individual-level behavior. Indeed, it highlights another not commonly discussed function of the social credit system: regulation of the economic activity of firms. While local social credit systems may not be used to monitor, constrict, govern, or even repress individual behavior, it *is* used to monitor, punish, reward, and govern firms.

What behaviors are punished?

If the targets of local social credit systems are firms, what are they targeted for? In other words, how are firms governed under the social credit system, and which behaviors are punished? One way to answer these questions is to examine the behaviors that attract the harshest punishments under the social credit system.

Three out of the four blacklists from Hangzhou, Fuzhou, and Hefei contain information that allowed me to code for severity of punishment. Hangzhou and Fuzhou's blacklists list the length of time the blacklist entry will remain on the blacklist. For Hangzhou blacklist entries, I coded severe punishment as remaining on the blacklist for 5 years or longer.[45] For Fuzhou blacklist entries, I coded severe punishment as remaining on the blacklist for longer than 79 years. This is mainly due to differences in how Hangzhou's and Fuzhou's blacklists report

[44] The individuals on the Hangzhou blacklist all come from local courts for noncompliance of court orders in paying back debt.

[45] Hangzhou blacklist entries may remain on the blacklist for less than one year, one year, 2 years, 3 years, 5 years, or 10 years, or have no limit.

time. Hangzhou's blacklist denotes a certain number of years the firm is expected to remain on the blacklist. In Fuzhou, the length of time a firm remains on the blacklist is written as an "expiration date." In other words, a blacklist entry is given a specific end date. An extreme example is an entry on Fuzhou's blacklist with an expiration date of 2099. The coding for Hangzhou and Fuzhou differs drastically therefore due to underlying differences in the length of the punishment period on the two blacklists. To calculate the time an entry will remain on the blacklist for Fuzhou entries, I counted the number of days between the date the entry was placed on the blacklist to the date of expiration. There is also much more variation in the time that Fuzhou blacklist entries remain on that list. Table 6.2 displays the summary statistics for these two variables.

Table 6.2 Summary statistics of severe punishment, by city

Hangzhou: Time on blacklist	
Length of time	*# of Entries*
< 1 year	4
1 year	5
2 years	56
3 years	995
5 years	2284
10 years	1
No limit	169
Until action taken	515
Fuzhou: # of days on blacklist	
Minimum	0
Maximum	31,481 (≈ 86 years)
Mean	13,696 (≈ 38 years)
Median	1096 (≈ 3 years)
Hefei: Amount fined	
Minimum	0 RMB
Maximum	3,050,000 RMB (≈ 471,100 USD)
Mean	14,870 RMB (≈ 2,280 USD)
Median	10,000 RMB (≈ 1,545 USD)

The Hefei blacklist does not include information on how long an entry will remain on the blacklist, but rather lists the type of punishment. Firms with a blacklist entry may pay a fine, be given a warning, have their assets confiscated, have their license revoked, or may be ordered to stop commercial activity. The vast majority of blacklist entries are fined (99%). For entries that are fined, the Hefei blacklist also provides information on the amount of the fine. I coded severe punishment as a fine exceeding 30,000 RMB or $4,600 USD.

Figure 6.4 displays for each city the behavior or rule violation that received severe punishment and the number of blacklist entries that were punished for each designated reason. For Hefei, the vast majority of violations that resulted in a severe fine regard proper permits for vehicles. There were other violations that appeared on the Hefei blacklist but for ease of display, I only include violations that appear more than 20 times.

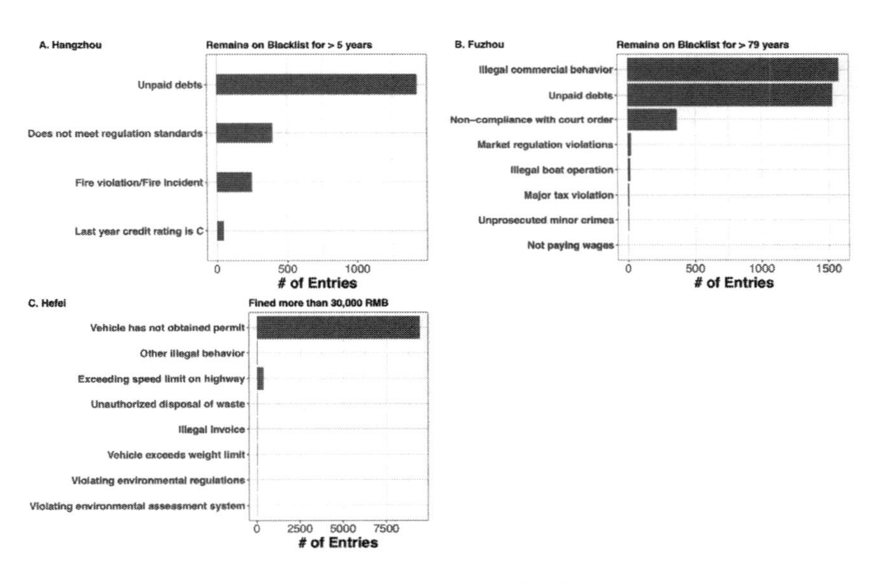

Figure 6.4: Violations that received severe punishments, by city

These analyses reveal a few insights about the types of behavior that attract harsh punishments and are therefore regulated under the social credit system. First, violations that result in a severe punishment are already illegal behaviors. Many of these violations concern unpaid debts or noncompliance with court orders, violations of market regulations or laws that govern commercial activity (such as not meeting certain regulation standards, illegal boat operations, or not paying workers' wages), violations that affect public safety or the environment (for example, fire code violations or incidents, violating environmental regulations, or unauthorized disposal of waste), and road safety (including not obtaining correct vehicle permits or exceeding speed or weight limits). The behaviors that local officials most want to deter through severe punishments are those that are already deemed illegal by existing laws and regulations.

Figure 6.4 allows us to examine the most common violations that receive harsh punishments. Examining these outliers can illuminate which behaviors are especially targeted under the social credit system. In the Hefei blacklist, there are 29 entries that received a fine of more than 500,000 RMB or approximately $77,200 USD.[46] These outliers are useful in understanding the kinds of behaviors that attract, albeit rarely, the most extreme punishment. Among these 29 blacklist entries, nine were violations of regulations concerning bidding on government contracts. Eight of the entries concerned failure to pass environmental pollution inspections or violation of other environmental regulations. Three of the entries were about work safety accidents that resulted in a death or severe injury that the firm then attempted to conceal. Other violations include selling unauthorized medical devices, illegal land occupation, violating construction permit regulations, selling fake food products, tax evasion, and fake advertising. These outlier entries attracting the harshest punishments (over $77,000 USD in fines) represent violations of regulations targeting commercial or economic activity and violations of regulations concerning public safety or the environment.

Examination of the reasons for harsh punishment under the social credit system reveals that the system targets the same behaviors as the existing legal system and regulatory agencies. Conversely, I do not find evidence that the social credit system is used primarily as a tool of surveillance and targeted repression; the data does not show that the behaviors targeted most harshly by the blacklists are political, anti-regime, or collective action behaviors. While we cannot observe cases where the social credit system is used for repression but is unreported, analyses of cases where the social credit system imposes quite harsh punishments provide evidence that the social credit system is used to secure compliance and deter violators of existing laws, regulations, and rules.

[46] These violations were not included in Figure 6.4 because they appeared on the Hefei blacklist less than 20 times.

Withering Surveillance and Repression?

Numerous commentators and recent academic papers have expressed concern over the surveillance and repression potential of the social credit system, concern that is specifically of interest to activists, lawyers, and journalists (Hoffman 2017; Xu et al. 2020; Liang et al. 2018; Creemers 2018). In addition, even if the social credit system is not currently used primarily for repressive purposes, there is little accountability or oversight in how the Chinese state implements the social credit system; hence, there always exists the potential for the state to leverage the social credit system as a coercive and surveillance apparatus.

The data collected for this chapter cannot definitively answer the question of whether the social credit system is used for surveillance in practice. The social credit system may be used for surveillance purposes in ways hidden to the public and researchers, and thus be very difficult to detect with existing data, and making it very difficult for citizens to know when surveillance is occurring. Using available data, however, I am able to examine the contexts where indirect evidence of repression and surveillance may be most likely to appear. If surveillance was occurring on a large-scale through the social credit system, we should at least be able to see evidence of it in the most-likely-to-be-detected cases, even if surveillance is difficult to capture in general.

Contributors to blacklists

If the social credit system is implemented with surveillance and targeted repression in mind, we might expect that information-gathering would be done by the coercive apparatus within local governments; however, if the social credit system collects and consolidates information across government agencies with the goal of increasing enforcement of laws, multiple agencies outside the coercive apparatus should participate in local implementation.

We learn two important facts by identifying who submits blacklist entries. First, data from local blacklists confirm that various agencies other than the public security bureau submit entries to blacklists. There are over 20 different government agencies that submit blacklist entries to the central blacklist. These other agencies dominate submissions to the blacklists; other than traffic police, the public security bureau does not actually submit any entries to local blacklists. Second, the fact that a large number of different government agencies submit to the blacklists illustrates the social credit system's role as an information-gathering and sharing apparatus. Information collected by individual government bureaus is then consolidated and shared across government agencies through one local blacklist.

Finally, the nature of government agencies that submit blacklist entries can reveal information about the kinds of behaviors that are targeted. I group government agencies into six categories: urban maintenance and public utilities, traffic, social services, regulatory agencies, the legal system, and government finance. I present in Figure 6.5 the total number of blacklist entries that are submitted by each type of government agency category across Hangzhou, Fuzhou, and Zhuhai.[47] Government agencies related to the legal system, regulatory agencies, and government finance submit the highest number of blacklist entries across the three cities. This suggests that the intended targets of the social credit system are individuals and firms that violate existing laws (such as those requiring compliance with court orders) and rules (regulatory violations, for instance). Local level blacklists are used to regulate markets, catch violators of low-level crimes, and assist legal institutions, highlighting again that the primary function of the social credit system is to reinforce existing legal and regulatory institutions and rules.

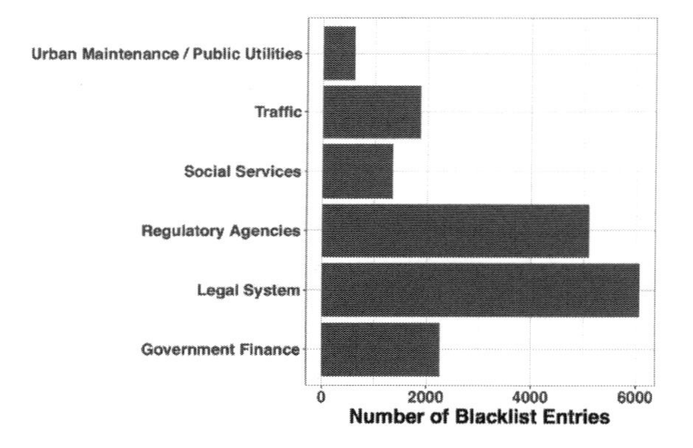

Figure 6.5: Contributors to blacklists

Personal social credit scores

As mentioned previously, certain pilot cities have implemented social credit scoring systems aimed at individuals, such systems having the greatest potential to be used as a repressive tool; however, analysis of the data available on these personal social credit rating systems also supports the claim that the social credit system is primarily used to increase compliance. Among the 16 pilot cities that have implemented a personal social credit scoring system, Fuzhou and Yiwu

[47] Almost all (91%) of the blacklist entries in Hefei are submitted by traffic agencies and so are not displayed in Figure 5 for ease of presentation.

have released detailed documents that list specific rules for subtracting or adding points to individual social credit scores. Previous commentary on the social credit system has emphasized personal social credit scoring systems as a potential tool for personalized repression or control. If the social credit system is intended mainly for repressive purposes, we may expect to see the rules for subtracting from social credit scores to use vague language that would offer greater opportunity for state organs to arbitrarily interpret the rules and use them to repress citizens. On the other hand, if the rules for subtracting from social credit scores make reference to specific regulations and laws, it can be much more difficult for local governments to apply them arbitrarily.

I coded all of the rules that pertain to subtracting from personal social credit scores from Fuzhou and Yiwu as being either explicitly connected to existing regulations, laws, or policies, or not. There is a total of 232 rules across the two cities, with Fuzhou listing 69 rules and Yiwu listing 163. For Yiwu, 85% (138 out of 163) of the rules resulting in decreases to a person's social credit score were related to an existing regulation, law, or policy. In Fuzhou City, 94% (65 out of 69) of the rules were related to an existing regulation, law, or policy.

Moreover, closer reading of the rules governing personal social credit systems provides additional evidence that the social credit system acts as a reinforcing institution, created to contend with weak enforcement of laws. An individual in Yiwu can lose points for selling and buying fake drivers' licenses or for submitting false information when applying to send their children to a school. Individuals who are the legal representatives of firms also lose points if their firms' social credit ratings decrease. Other reasons include violating fire protection laws, breaking traffic laws, or tax evasion. Similarly, in Fuzhou, an individual who does not comply with administrative penalties (such as paying a fine) for illegal behavior will suffer a decrease in their score of 30 points. If an individual violates an existing law and obstructs attempts to enforce administrative penalties using violence and threats, they will lose 150 social credit points. These examples demonstrate that the personal social credit system is used as an additional punishment mechanism when implementation of administrative penalties proves difficult or inadequate.

While the overwhelming majority of rules that govern personal social credit scores pertain to already existing laws, regulations, and policies, we do observe how a small minority of rules under this system have the potential to be used for repressive purposes. In Yiwu, for example, petitioners whose complaint is determined to be "false, fabricated, distorted, or misrepresented" can lose 6 social credit points. At times, citizens use petitioning to report local government misdeeds or even local corruption, in the hope that upper-level governments will intervene on their behalf (Cai 2004; 2010). In such cases, local governments have an incentive to stop petitioners. By including "false" petitioning as one of the factors that decrease one's social credit score, the local government could punish political petitioners using the personal social credit system—but this special case is an exception, not the rule. As I have documented above, the vast majority of rules that determine one's social credit score are tightly tied to existing laws, regulations, and policies.

What is the true function of the social credit system? How does it contribute to local governance? A close examination of local social credit systems reveals that their local manifestations reinforce compliance with laws and regulations in support of existing legal and regulatory institutions. Specifically, firms—not individuals—are the primary targets of local social credit systems. In addition, social credit systems regulate and punish violations of existing rules and regulations, including violations of regulations relating to bids on government contracts, violations of environmental regulations, or violations targeted at selling fake products.

Though there has been much political discourse around the social credit system, particularly in Western nations, such discourse has emphasized the surveillance and repression potential of the social credit system. While the evidence presented in this chapter cannot definitively rule out that repression or surveillance may be taking place, the available data reveals an equally important and verified function of the social credit system, the reinforcement of existing legal and regulatory mechanisms. The punishable offences under the social credit system are all violations of existing regulations of commercial activity, public safety, or the environment. Finally, even in pilot cities that do target individuals, personal social credit scoring rules pertain to already existing laws, regulations, and policies. It is clear that the social credit system does not exist to police or constrain new types of behavior. In fact, it acts to reinforce an existing set of rules, regulations, and laws.

These findings do not definitely prove that surveillance or repression is not happening through the social credit system. Surveillance and repression are, by their nature, difficult to detect or measure. Governments want to hide their engagement in these activities, making it difficult for researchers to access relevant data. The data and evidence provided in this chapter, however, does suggest that the social credit system may have other functions beyond surveillance. The finding that many social credit systems target firms rather than individuals, and that local social credit systems mete out the harshest punishments to violators of commercial and public safety regulations, highlights the regulatory and legal function of the social credit system—a point often overlooked by depictions of the social credit system as an Orwellian tool of social control.

Conclusion

This chapter has sought to answer a fundamental question about the social credit system: what are its political uses and functions? Drawing upon local government documents, an original coding of implementation efforts across 43 cities, and actual records of punishments under local social credit systems, I have demonstrated that one overlooked function of the social credit system is its law-enforcing mechanism. This claim is buttressed by two crucial findings: that firms are the major targets of local social credit systems, and that they are largely punished for violating existing laws and regulations. While the social credit system's potential as a tool for surveillance and repression cannot be definitively rejected, the available evidence points to an equally significant function of the social credit system. It is used to buttress and support existing legal and regulatory frameworks, acting as an additional enforcement mechanism to ensure the compliance of market actors.

Beyond the specific findings about the social credit system's governance functions, this chapter has highlighted the need to examine the evolution of the social credit system at local levels. Similar to other work in Chinese politics that emphasizes the large impact of local political actors and processes on important political outcomes of interest, this chapter localized the social credit system. By studying the outputs of local governments' social credit systems, we can more systematically map how the social credit system works in specific contexts, which is the foundation for theorizing about its consequences and effects.

References

Ang, Yuen Yuen. 2024. "Ambiguity and Clarity in China's Adaptive Policy Communication." *The China Quarterly* (257).

Brussee, Vincent. 2022. "China's Social Credit Score – Untangling Myth from Reality." *Mercator Institute for China Studies.* https://merics.org/en/comment/chinas-social-credit-score-untangling-myth-reality. Accessed October 1, 2024.

Cai, Yongshun. 2004. "Managed Participation in China." *Political Science Quarterly* 119 (3).

Cai, Yongshun. 2010. *Collective Resistance in China: Why Popular Protests Succeed or Fail.* Stanford: Stanford University Press.

Carney, Matthew. 2018. "Leave No Dark Corner." *Australian Broadcasting Corporation*, September 17, 2018.

Chung, Jae Ho. 1995. "Studies of Central-Provincial Relations in the People's Republic of China: A Mid-Term Appraisal." *The China Quarterly* (142).

Creemers, Rogier. 2018. "China's Social Credit System: An Evolving Practice of Control." Working Paper.

Economy, Elizabeth. 2018. *The Third Revolution: Xi Jinping and the New Chinese State.* New York: Oxford University Press.

Engelmann, Severin, Mo Chen, Felix Fischer, Ching-yu Kao and Jens Grossklags. 2019. "Clear Sanctions, Vague Rewards: How China's Social Credit System Currently Defines Good and Bad Behavior." In Proceedings of the Conference on Fairness, Accountability, and Transparency (FAT* '19). Association for Computing Machinery, New York, NY, USA, 69–78.

Fearon, James and David Laitin. 1996. "Explaining Interethnic Cooperation." *American Political Science Review* 90 (4).

Heffer, Abbey S. and Gunter Schubert. 2023. "Policy Experimentation under Pressure in Contemporary China." *The China Quarterly* (253).

Hoffman, Samantha. 2017. "Programming China: the Communist Party's Automatic Approach to Managing State Security." PhD diss., University of Nottingham.

Kim, Jinwoo. 2017. "Orwell's Nightmare: China's Social Credit System." *Asian Institute for Policy Studies*, February 28, 2017.

Kostka, Genia and Lukas Antoine. 2020. "Fostering Model Citizenship: Behavioral Responses to China's Emerging Social Credit Systems." *Policy & Internet* 12 (3).

Liang, Fan, Vishnupriya Das, Nadia Kostyuk and Muzammil Hussain. 2018. "Constructing a Data-Driven Society: China's Social Credit System as a State Surveillance Infrastructure." *Policy & Internet* 10 (4).

Liu, Lizhi. 2018. "From Click to Boom: The Political Economy of E-Commerce in China." PhD diss., Stanford University.

Mercator Institute for China Studies. 2021. "China's Social Credit System in 2021: From Fragmentation Towards Integration."

Montinola, Gabriella, Yingyi Qian and Barry Weingast. 1995. "Federalism, Chinese Style: The Political Basis for Economic Success in China." *World Politics* 48 (1).

Mozur, Paul. 2019. "China Moves Towards 'Digital Totalitarian State' As Surveillance Technology Continues to Advance." *The Independent*, December 18, 2019.

Oi, Jean. 1999. *Rural China Takes Off: Institutional Foundations of Economic Reform.* Berkeley: University of California Press.

Oi, Jean. 2011. "Politics in China's Corporate Restructuring." In *Going Private in China: The Politics of Corporate Restructuring and System Reform*, ed. Jean Oi. Stanford, CA: Walter H. Shorenstein Asia-Pacific Research Center Books.

Shirk, Susan. 2018. "The Return to Personalistic Rule." *Journal of Democracy* 29 (2).

Takeuchi, Hiroki. 2014. *Tax Reform in Rural China: Revenue, Resistance, and Authoritarian Rule.* New York: Cambridge University Press.

Tsai, Lily. 2007. *Accountability Without Democracy: Solidary Groups and Public Goods Provision in Rural China.* New York: Cambridge University Press.

Xu, Xu, Genia Kostka and Xun Cao. 2022. "Information Control and Public Support for Social Credit Systems in China." *Journal of Politics* 84 (4).

CHAPTER 7

Scientific Fairness: Justifications and Critiques of Points Systems in Shenzhen

Anne-Christine Trémon

Abstract

Points systems feature centrally in China's embrace of "smart governance" and e-government. These systems allocate rights to urban citizenship or to locally-provisioned public goods on the basis of a numerical assessment and ranking of applicants. In Shenzhen, one of China's largest cities with the highest proportion of migrants in its population, municipal authorities justify the use of points systems by emphasizing the good governance principles they can put into practice by using them. This chapter analyzes the justifications that designate points systems as a scientific governance tool that ensures fairness. This discursive legitimation invokes not only their scientific (*kexue*) but also their "reasonable" (*heli*) character. This shift from "scientific" to "reasonable" makes "scientific fairness" understandable to the public, and opens up a space for potential critique; however, the justifying adjectives of "scientific" and "reasonable" act both as tools of critique, and as tools for discouraging critique.

Keywords: Points systems; Hukou; Public goods; Fairness; Selection; Evaluation; Automated governance; Open government; Shenzhen

Introduction

China's points-based systems reflect the worldwide trend toward digitized and automated governance. While modern state projects of social engineering have from early on developed quantification and statistical tools (Hacking 1991; Rose 1991; Scott 1998), quantitative indicators are now increasingly used to audit organizations, evaluate performance, and enforce good governance (Rose and Miller 1992; Strathern 2000; Desrosières 2010; Shore and Wright 2015). Big data collection and data-driven modeling have further fueled the development of automated decision-making processes, which several authors warn are based on abstract numbers that are taken as fact, amount to depoliticized adjustment mechanisms, and render policies immune to debate (Supiot 2015; O'Neill 2016; Zuboff 2019).

China has heavily invested in "smart governance" technologies based on automated decision-making to improve legal and regulatory procedures, as well as investing in e-government solutions to increase openness and transparency. A growing body of literature is concerned with the implications of digitalization and automation for the management of cities, education, healthcare, financial reporting, and the judicial system, with special attention being paid to the question of whether China's social credit project prefigures a total surveillance society (Daum 2017; Creemers 2018; Zhang 2020). This literature partly overlaps with that on e-government, although this strand focuses rather on the extent to which information technologies enhance the possibilities of public participation and policy legitimation under authoritarian governance (Chen, Pan and Xu 2016; Distelhorst and Hou 2017; Gueorguiev 2021; Chen et al. 2023).

Points systems (*jifen zhidu*) feature centrally in China's move toward "smart governance" and e-government. They have become generalized in China's larger cities, mostly as a means of granting *hukou*. *Hukou* is a form of social citizenship since it determines access to welfare benefits and public services (Solinger 1999; Smart and Smart 2001; Liu 2008, Woodman and Guo 2017). Points systems were first experimented with in the Pearl River Delta cities of Guangdong province in 2010. Following the central government's reform of the household registration system in 2014, these systems have become mandatory in Chinese megacities, i.e. cities with over 5 million urban hukou holders (State Council 2014a).[48] Similar points systems have been widely adopted for extending access to public goods that were previously restricted to *hukou* holders, mainly public school places. In this way, points systems allocate rights to urban citizens or allocate locally-provided public goods on the basis of a numerical assessment and ranking of applicants, and in accordance with quotas or the number of available places.

Points systems readily lend themselves to digitalization and thus automation. I use the term digitalization rather than digitization since points systems are, per definition, numerical. China has a well-established tradition of points-based evaluation of work performance, as practiced in Mao-era agricultural production teams and within urban work units, with that tradition extending to each individual's moral conduct starting from secondary school (Unger 1984; Bakken 2000). Numerical assessment methods are meant to reduce the arbitrary judgments associated with human intervention. Over the past twelve years, automated decision-making has been fostered by the conversion from paper to electronic application files, the construction of digital databases, and the use of information technologies to communicate policies, process applications, and allow for public feedback.

[48] The National Bureau of Statistics classifies cities with over 5 million urban inhabitants as megacities (teda chengshi) and cities with over 10 million urban inhabitants as super-large cities (chaoda chengshi). In 2021, there were 14 megacities and 7 super-large cities respectively (National Bureau of Statistics 2021).

Points systems are a flexible tool for allocating public goods either by way of *hukou* or exclusive of it. In China's larger cities that have the most generous public goods regimes, *hukou* is particularly hard to obtain. Points systems filter *hukou* grants by ranking applicants according to their total score, calculated by adding up points corresponding to such criteria as age, degrees and skills, social security payments, conformity of behavior, and investment capacity. They ensure that an increase in the registered population will not result in greater pressure on the local budget (Zhang 2012; Zhang and Li 2016). Points systems also allocate available places in public schools to non-priority non-*hukou* holders on the basis of quantified criteria such as home ownership and length of contribution to urban social security. Points systems thus stratify citizenship (Li, Li and Chen 2010; Guo and Liang 2018; Dong and Goodburn 2020; Zhang 2018) by differentially distributing rights to public goods according to legal status and economic and cultural capital.

As a technology used to implement China's central and municipal plans of population distribution and resource allocation, points systems play a pivotal role in population management. Some scholars have taken inspiration from the Foucauldian concept of governmentality to argue that the *hukou* system at large (Wang and Liu 2018), and points systems in particular (Zhang 2018), represent a form of authoritarian governmentality. Elsewhere (Trémon 2025) I show that China's numerical assessment techniques for regulating its population encompass the basic components of Foucauldian governmentality. Since their primary purpose is fostering the welfare of the "population", they use the science of political economy to calculate its optimal distribution, and they resort to disciplinary techniques to regulate its flow and behavior (Foucault 1991, 102).

However, governmentality approaches often neglect the communicative dimensions of legitimation and justification (Barnett et al. 2008; Kipnis 2007; Watts 2022). Justification is crucial because points systems are highly complex, constantly evolving, and extremely selective. The establishment of points systems has accompanied significant pedagogical efforts in policy documents, online guidelines, and platforms allowing the public to post queries and suggestions. Points systems are a form of exercise of state power that is highly visible, and whose game rules are made known to all participants. There is an element of sovereign spectacle in the campaign-like functioning of points systems, characterized by their publicly announced yearly opening dates that are actively advertised. Moreover, the discourse that explains points systems to the public resorts to common sense morality. The main idea is that only deserving individuals obtain public goods, that is, those who have invested in the city through their work and contributions. In this regard, points systems pertain to what Bakken has termed, an "exemplary society", one that "leans heavily on mechanisms of morality and education" (2000: 5, 100).

This chapter therefore fills a gap in the literature by examining how, in Shenzhen, one of China's largest cities with the highest proportion of migrants in its population, municipal authorities justify the use of points systems by emphasizing the good governance principles such systems allow them to put into practice. The concept of "scientific fairness", aired in a municipal policy paper (Shenzhen 2021), nicely encapsulates how points systems are generally presented: as a scientific method ensuring fairness.[49] Science featured prominently in China's goals of modernization and economic growth in the reform era after the chaos of the Cultural Revolution, and it gained in importance following the global economic crisis in 2008 which affected its export-driven economy. Scientist and solutionist attitudes toward science and technology prevail, imbued with a mix of Marxist-Leninist notions of laws ruling society and nature, and cybernetics and systems engineering (Bakken 2000; Greenhalgh 2003; Sigley 2006; Greenhalgh and Zhang 2020). This belief in the power of science grounds the Party-State's continued practice of planning as well as its emphasis on the social engineering (*shehui gongcheng*) of individual conduct.

The claims to scientific fairness legitimate points systems. Due to the fragility of those claims, however, they also constitute the limits of their effectiveness and of people's acceptance. This chapter uses a combination of critical discourse analysis (CDA) and rhetorical analysis (Fairclough 2013; Goddard and Krebs 2015) to show how, when associated with fairness in policy discourse on points systems, "scientificness" takes on much less scientist overtones. As the case of Shenzhen illustrates, their legitimation rests on the invocation not only of their scientific (*kexue*) but also, and almost as frequently, their "reasonable" (*heli*) character. The shift from "scientific" to "reasonable" makes "scientific fairness" understandable to the public; however, it also opens up a space for potential critique, since the extent to which a policy is "reasonable" can be judged and debated. Because of the self-referential functioning of points systems (Luhmann 1995), the justifying adjectives of "scientific" and "reasonable" act both as tools of critique, and as tools for shutting down critique.

I conducted fieldwork between 2011 and 2018 in an urban village in Shenzhen's northern Longhua district which hosts a large population of migrants. Although the research project at that time did not initially focus on points systems, their importance emerged during interviews with migrant residents. Due to the inaccessibility of the field after 2020, I made the decision to conduct online research instead. This chapter analyzes online policy documents and netizens' postings on governmental and commercial forums in Shenzhen over the period 2010 to 2022. The data collected between April and September 2022 consists of 60 policy

[49] Only one recent article on points systems for allocating school places (Wan and Vickers 2022) notes that their main justifying principles are equity and transparency.

documents (mostly at city level) published on government websites, 60 articles published by commercial counseling agencies (including 28 followed by online discussions), 18 articles published by NGOs or independent bloggers, all mainly on WeChat public accounts, Zhihu question-and-answer forums, Baidu Baike online encyclopedia, 28 posts and responses on message boards for leaders, an official municipal channel for asking the government questions and expressing discontent mostly about public services, 7 articles published in state media, and 29 court case reports (not used in this chapter) which I coded using Hyperresearch software.

The first section outlines how points systems have developed in Shenzhen since 2010 as part of the city's policy of economic upgrading in response to China's "scientific development" goals and urbanization plans. The second section unpacks the meanings of the scientific character attributed to points systems, and analyzes the ways in which such systems are justified by being presented as "reasonable". The final section shows how citizens' online critiques of points systems all reflect a lack of information and uncertainty about their operation, critiques that are at odds with the government's claims to open and reasonable population governance. The chapter concludes with this paradox: points systems' proclaimed "scientificalness" and fairness opens them to critiques from the public, but also stifles such critiques on these same grounds.

Scientific Development and Point Systems

From early on, Shenzhen's urban population management policies were placed under the aegis of the 'Scientific outlook on development' (*kexue fazhan guan)* concept. This section focuses on the origin of "scientific fairness", that is, "scientific development", a guiding concept of China's social and economic policy since the early 2000s, a policy in which points systems are a major instrument. "Scientific outlook on development" made its first appearance in a speech by president Hu Jintao in November 2002, then was written into the Chinese constitution in 2018. It meant that China had to change the course of its development path from fast growth over-reliant on a cheap labor force and the overuse of natural resources to a more qualitative development based on well-educated workers and environmental preservation, from a largely export-based industrial economy to one in which high-tech industry and the tertiary sector played a larger role driven by domestic demand, and from the assessment of economic performance based on GDP indicators to the assessment of people's wellbeing through happiness indicators (China Daily 2010).

In the official discourse, a "scientific outlook on development" continues to underpin economic development, but the government must nevertheless be people-oriented (*yi ren wei ben*) and construct a harmonious society, ensuring prosperity for all, fairness and justice, and the harmonious coexistence of humans and nature. The need for coordinating these different aspects of development means that the Party plays a leading role in "conducting unified planning" (China.org 2021). In the Chinese version of orthodox Marxist thought, the development of society can be explained by objective laws, knowledge of which enables scientifically-planned governance (Creemers 2020); thus, the Party-state should remain the primary driving force behind national development aimed at reaching the ultimate state of "ecological civilization" (a concept also coined by Hu Jintao), that is, a green, high-tech and urbanized society (Rodenbiker 2021; Trémon 2022). Although science was the first of Deng Xiaoping's reform-era's 'four modernizations', it has gained in importance as a way of upgrading China to be an "innovation nation" (Greenhalgh and Li 2020).

The New Urbanization Plan (NUP) applied the scientific development concept to urban development between 2014 and 2020 (State Council 2014a). The NUP stated that the model followed in the 1990s, of rapid urbanization relying on cheap labor supply and unequal public services, was unsustainable; rather, urban public service coverage, and therefore the rate of urban *hukou* holders should increase. It hailed urbanization as the main path to modernization, while stressing that the urbanization of the Chinese population should be realized "rationally" or "reasonably" (*heli*). The Plan also asserted that while development should be driven by the market it must be supervised by the government. While the government retained its planning role, such planning should adjust to the needs of the market. Population management had therefore to ensure that human resources matched the needs of employers and offered a favorable institutional setting that would allow the economy to grow. This is illustrated by the points system in Shenzhen.

Shenzhen's points system was introduced in 2010 as part of a regional experiment that was extended nationally by the NUP. It made points systems mandatory in cities with a population above five million. From the early 2000s, Shenzhen, a new city which had developed as a Special Economic Zone based on export industries (Sklair 1991; O'Donnell, Bach, and Wong 2017), sought to upgrade its economy by encouraging investment in higher-tech, greener, and less labor-consuming industries and services. The 2005 *Opinions on strengthening and improving population governance* announced that the city's population policy should be "guided by scientific development" and follow "the objective laws of urban development and population development" (Shenzhen 2005). The city's demographic evolution, in both size and composition, should allow for this staged development to unfold (article 1), while the threshold for *hukou* registration should be scientifically determined according to the needs of economic and social development (article 2).

Shenzhen accelerated its economic conversion after it was badly hit by the 2008 global economic crisis. Reflecting the Twelfth National Five-Year Development Plan of Science and Technology (2011–2015), Shenzhen's twelfth Five-Year Plan, launched in 2011, was meant to open up "a new path of scientific development" substituting "Shenzhen quality" (*zhiliang*) for "Shenzhen speed" (*sudu*) (Shenzhen 2011).[50] Shenzhen increased its promotion of a series of policies targeting innovative talents (*chuangxin rencai*). These were a continuation of its earlier blue-chop (*lanyin*) *hukou* policy, introduced in 1995. Blue chop *hukou* status offered most of the advantages of permanent registered *hukou*, and could be obtained either by buying a commodity flat or by fulfilling certain age and employment conditions. Admission to the latter category, subject to an annual quota set by the municipal public security bureau, relied on a points-based assessment system. Blue chop personnel were ranked and selected according to the total number of points scored for educational level, age, time spent working in Shenzhen, occupational title, profession, and honors (Wong and Po 1998). This early points-based assessment method was suspended in 2005 when Shenzhen replaced its blue-chop policy with a talent attraction policy. Framed by the 2005 *Opinions*, this policy targeted overseas returnees, as well as high-level professionals and entrepreneurial investors. Moreover, there was no numerical assessment system and no quota. Applicants could be directly granted Shenzhen *hukou* status provided they fulfilled certain age criteria, which varied along with level of degree, or skills, and amount of taxes paid.

Meanwhile, two Opinions issued by the State Council in 2006 and 2011 marked a change in awareness of the central authorities regarding the stark inequalities generated by the *hukou* system (State Council 2006, 2012). In this context, the Guangdong provincial government issued the *Guidelines for Carrying Out Work on Peasant Workers' Points-based Access to Urban Hukou* in 2010, requiring Guangdong's largest immigrant cities, all located in the Pearl River Delta, to implement points-based access to urban *hukou* (Guangdong 2010; Wang 2020). The first official points-based channel opened in 2010 shortly after Shenzhen introduced a new ten-year residential permit which accorded some basic rights to migrant residents and constituted a precondition for applying for *hukou*. The new channel was tailored for a category of people that were not included in the select categories mentioned above. It extended the possibility of *hukou* acquisition in the name of promoting the "citizenization" (*shiminhua*) of migrant workers, and was based on an explicit logic of merit according to which people who had long made contributions to the cities in which they lived should be granted *hukou*.

Between 2010 and 2016, two ways of acquiring *hukou* coexisted: direct approval (*hezhun*) of "talents" provided they met certain criteria, and points-based selection

[50] "Shenzhen speed" reflects the earlier phase of Shenzhen's development as a Special Economic Zone and as a newly-created city that was built at an amazing pace and whose population growth challenged all expectations.

of migrant workers within the limits of an annual quota. Soon after the first trials, however, the score indicators gradually tightened, reflecting a decrease in the annual quota. It is notable that labeling the policy "trial" was not so much intended to acknowledge that a pilot policy was being tried out, but that the policies were constantly changing in response to a need to adjust the flow and structure of the population almost in real time. All policy measures related to points systems over the period 2010 to 2021 were labeled "trial". This allowed the city's authorities to keep flexible control over the population according to the economy's needs and changing budgetary means.

Points systems were designed as technical tools for accomplishing municipal population and economic development goals, and thus to assist in accomplishing science- and technology-led upgrading. Points systems' policies have, from the start, emphasized the importance of strengthening the "macro-control of the city's population" and accomplishing "population transformation through industrial transformation", both aimed at the "comprehensive and coordinated development of population, economy and society" (Shenzhen 2010). As early as 2010, the city used academic or skills requirements, called indicators of "quality" (*suzhi*), in its index for calculating the applicants' scores, thereby selecting the most desirable new *hukou*-holders among those who had been living in the city for a long time.

Although the newly-created points-based channel was flagged as a means of extending *hukou* access to long-term residents who had no hope of accessing *hukou* any other way because of their low cultural and economic capital, from the start it selected people under a certain age who enjoyed a stable work contract measured by their years of contribution to social insurance. It also selected people according to their "quality" as measured by their education (ranging from 5 points for middle school education to 60 points for a PhD in the 2010 index table) and by bonus criteria which yielded additional points, such as participation in blood donations and volunteer services.

These practices temporarily stalled in 2016 after the State Council released the National New Urbanization Plan (2014–2020) and *Opinions on Further Promoting the Reform of the Household Registration System* (State Council 2014 a, b). Cities that had a low proportion of *hukou* holders were to actively contribute to one of the NUP's most prominent goals, namely the granting of *hukou* to approximately 100 million people by 2020. Shenzhen is the Chinese city that had (and still has) the largest share of non-*hukou*-holding people in its migrant population due to its development as a Special Economic Zone. At the end of 2015 the population of Shenzhen had reached 11,378 million, only 3.54 million (31.2%) of whom were *hukou* holders (Shenzhen Statistics Bureau 2016). Shenzhen's 13th Five-Year Plan proposed that by 2020 the city's resident population target be set at 14.8 million, 5.5 million of whom would be *hukou*-holders (Shenzhen 2017a).

To meet this goal, in 2017 Shenzhen adopted a new "talent introduction method", dividing the points system into two channels for obtaining *hukou*. The first was called "pure points", for applicants under the age of 55. There were no longer any academic or skills requirements, points accruing only according to the number of months of participation in Shenzhen social insurance and time owning or renting a house. In parallel, however, the other points channel that assessed academic degrees and professional certificates targeted those who did not meet the age conditions for direct approval of talents. Both channels competed for places within the quotas for points-based admissions, quotas which rose to 10,000 persons per year in 2017, 2018, and 2019. In parallel, the number of people admitted to Shenzhen citizenship through simple approval exceeded 230,000 in 2017, 237,000 in 2018, and 280,000 in 2019.

Shenzhen's proportion of *hukou*-holders still stood at only 29% in 2020. Although this was the outcome of a continuous immigration inflow, the figure indicates that, like other Chinese cities participating in the "talent race" (Shen and Li 2020), Shenzhen paid lip service to the central authorities' *hukou* reform, mainly recruiting talents (see Trémon et al. 2024). After the NUP 2014–2020 expired, the new policy, announced in 2021, much less ambiguously extended the logic of points-based assessment and quota-limited admissions to most categories of applicants, creating a single system intended first and foremost to attract highly educated and skilled people (Shenzhen 2021).

Points System as the "Reasonable" Method

While population management is a vector of scientific development, it is also the management method itself which must be scientific. The points system is defined in policy texts as a scientific method due to its recourse to a quantified scoring index. The very first Trial measures therefore stated that "The point-based household registration method refers to the *scientific* establishment and determination of a point-based index system, and assignment of a certain score to each index, to *quantify* the conditions for migrant workers to access Shenzhen hukou" (Shenzhen 2010, article 2). This reflects the prevailing scientist attitude that equates quantification with objectivity (Thornton 2011: 250; Greenhalgh 2003); however, in policy texts and announcements, the adjective *heli*, which means "rational" and "reasonable", often joins or even replaces "scientific" (*kexue*). The 2012 Trial measures introduced the expression "entering the household with points" (*jifen ru hu*), defined as the "multi-evaluation and comprehensive scoring of migrant workers" applying for *hukou* admission "through the establishment of a *scientific and reasonable* quantitative index system" (Shenzhen 2012, article 4). Henceforth, the quantitative index system was systematically characterized as both scientific and reasonable. *Heli* has a more subjective connotation than *kexue*; it pertains to everyday language and therefore appeals to a wider moral sensibility among the public. It changes the language register from formal and abstract to familiar and concrete, and it shifts the emphasis from science to policymaking, suggesting that the policy is the best option under given circumstances. This section unpacks the five ways in which points systems are presented as "reasonable".

First, points systems are scientific and reasonable because they rely on calculations and projections by the hard sciences. One example is the NUP stipulating that cities should improve "scientific" urban planning by calculating cities' "carrying capacities" (*chengzai nengli*) and planning their space and population growth accordingly in order to avoid overuse of land and inadequate infrastructure (State Council 2014a). "Carrying capacity" is a Malthusian concept coined by Club-of-Rome theorists who, in the late 1970s, influenced the Chinese scientists that recommended the one-child policy. They adopted a Malthusian stance that had been rejected under Mao due to Marx's negative critique of Malthus (Greenhalgh and Winckler 2005). In requiring that China's largest cities (over five million) strictly control their population size and improve or establish points systems to achieve this, the NUP continued these politics of population at a city-level scale. The same population scientists were influenced by cybernetics, its focus on systems (*xitong* or *zhidu*) being central to municipal population plans and points system policies.

In central and municipal policy documents, the technical term "carrying capacity" is often mentioned, but is also translated into more accessible terms, with recourse to the adjective *heli* to describe a way of planning that avoids any imbalance between population and resources. In the New Urbanization Plan of 2014, the 46 occurrences of the word *heli* are used in this sense. Likewise, in Shenzhen policy texts, points systems aim to redress the "unreasonable population structure" and "reasonably control the total population size" (Shenzhen 2005). It is often explained, especially in policy documents relating to points systems for allocating school places, that the city's resources are limited, and should therefore be reasonably distributed. "Reasonable" invokes a rationale of scarce resources that is of a higher order (because it is a matter of calculus) and justifies the fact that rights to education, for instance, are made conditional upon available resources. In this way the points systems are justified by their *raison d'être,* that is, the allocation of scarce resources.[51]

Second, this allocation is considered reasonable in that it preserves not only the quality of public goods by adjusting the population to available resources, but

[51] For further reflection on this point and on "carrying capacity" as central to spatial governmentality see Trémon 2025.

also the quality of the population. Points systems are not based on a notion of rights granted unconditionally to citizens once they fulfil certain criteria, but on conditional rights, which should be earned individually, by those who have earned merits.[52] Central-level policy documents state that urban citizenship should be extended to people "*capable* of stable employment and living in cities and towns" (State Council 2014b); therefore, worthy recipients of urban citizenship or school places are individuals of high "human quality" as measured in the points system indexes (see previous section). Points systems are openly presented and justified as conducive to "optimizing" (*youhua*) the city's population quality. The Chinese discourse on human quality (*suzhi*) has a strong pedagogical intention (Bakken 2000). The points systems encourage temporary residents who hope to obtain *hukou* to plan early and make sacrifices, by buying social security locally rather than saving to eventually build a house in their home town, and by spending money on evening and weekend courses. They incentivize people to start buying social security well ahead of time so as to secure *hukou* by the time their child reaches school age. They have given impetus to a flourishing cohort of private business agencies selling *hukou* acquisition packages that include counseling and training for a degree (part-time undergraduate degrees yielded additional points until 2020).

The systems also encourage applicants to behave morally, but above all they discourage immoral or illegal behavior. In Shenzhen, the earlier points systems (until 2017) added bonus criteria such as participation in blood donations and volunteer services, and 30 bonus points are still added in case of commendations for bravery. The number of points subtracted in case of unlawful conduct amounts to reducing an individual's score to zero. Until 2022, points were subtracted mainly for noncompliance with the single child policy. The latest points system policy, which was announced in February 2023, deducts 200 points in cases of negative behavior, and 300 points in case of illegal behavior. "Negative behavior" is recorded and verified by the Municipal market supervision administration, which has run Shenzhen's social credit system since March 1, 2023, the date when it came into force (Shenzhen MMSA 2023). The list of 15 types of "negative behavior" leading to point subtraction classifies them into three types: "acts that endanger the health and life safety of the people", "acts that undermine the fair competition order of the market and the normal order of society" and "acts that fail to perform legal obligations and affect the credibility of judicial and administrative organs".[53]

[52] Maoist thought proposed a correction to the Marxist position that members of the same class share the same dispositions, by insisting on the particularities of individual persons: new abilities and skills can be forged where they had not previously existed (Munro 1971: 630).

[53] This recent connection between hukou and school points systems and the social credit system accounts for why this topic is little explored in this chapter.

Third, points systems grant rights in the context of competition. The science of evaluation that has developed in China rests on the idea that feelings of stress arising from competition are a highly motivating force, producing in everyone a will to strive continuously (Bakken 2000, 62). The policy document that endorses the principle of "scientific fairness" associates it with another principle, that of "competition for the best" (Shenzhen 2021). The underlying rationale is that the more applicants there are for *hukou*, and the harder it is to obtain it, the better the population's quality. This further accounts for the practice of higher-levels of government constantly urging lower levels to implement *reasonable policies*. Indeed, such an injunction can be found in policy documents at central or provincial levels inviting local governments to set reasonable criteria, or among city governments, in the case of points systems, for accessing public school places: district educational bureaus are urged to set reasonable indexes. "Reasonable", in all these cases, implies that points systems should not set such high scores as to seem completely out of reach. Points systems work only if people believe they have a chance of beating the competition, and work best if many competitors participate in the race. Private agencies that report on the latest policies and make efforts at explaining how points systems work often compare them to the *gaokao*, the national competition for accessing universities.

Fourth, points systems are a scientific and reasonable *method*. Every single policy document that concerns points-based *hukou* access in Shenzhen starts by characterizing them as such. Here, quantitative is treated as synonymous with scientific and objective. The quantitative treatment of the applicants' data and their ranking are also presented as fair, that is, ensuring an equitable and impartial treatment of candidates. Impartiality is achieved through automatic comparison of applicants based on their scores; automation is meant to eliminate administrative arbitrariness by selecting the applicants according to their ranking. From early on, tight procedures based on detailed quantified indexes were meant to depersonalize bureaucratic intervention (Bakken 2000, 299), while the pivot to automation technology was driven by the idea that it would solve problems of arbitrariness and corruption within the bureaucracy. In every year since points systems were first implemented in 2010, all applicants have lined up according to the same standard because the calculation of their age and of their points on each indicator has been based on the score they reached on the same date (for instance, June 30). The application and processing procedures were digitalized in 2013. The processing time for applications, which was several months prior to digitalization, was reduced to one month.

Approval or rejection is determined by a score which can be calculated by the applicant themselves, and is thus fully automated; what takes time is the matching of information on the candidates' credentials by all the departments involved (social security, housing, employment, and so on). This has required the construction of a shared municipal database, meant to reduce intergovernmental information asymmetries, speed up processes, and limit fraud. Digitalization and arguments of scientific fairness have allowed the extension of the numerical assessment method to other categories of applicants, first to talents, making it harder for ordinary workers to compete, and then progressively to all categories of applicants, in the name of policy systematization and "competition for the best".

Finally, the Shenzhen municipal authorities put forward several good governance principles that guided their implementation of points systems, namely publicity, transparency and accountability. China's "Open Government Information Ordinance" (State Council 2007) required that government websites serve as key platforms of information disclosure and online responsiveness.[54] Following national regulations, Shenzhen emphasized the importance of building a legal, service-oriented, and responsible government (Shenzhen MHRSSB 2010). *Hukou* policies were then published and explained on city governments' websites, providing guidelines to help people navigate the different available channels according to their applicant category. In 2011 the Municipal Human Resources and Social Security Bureau (MHRSSB) released the first "Shenzhen Migrant Workers Points-based Operation Guidelines" on its website. Since then, numerous guidelines and online Q&A sessions have been posted on that website and on the WeChat account of the Public Security Bureau (PSB, which took over from the MHRSSB in 2016). These were necessary because different categories had been re-ordered, new channels created and then fused, and indexes changed numerous times, generating a constantly-evolving system, the complexity of which made it difficult to grasp.

The allocation of urban citizenship and school places was made open and transparent throughout the process from the announcement of the policies informing applicants about the criteria and rules for calculation, to the publication of results. Decisions had to be openly announced within a given timeframe. Article 10 of the 2011 Trial measures in Shenzhen stipulated that the lists of names of approved candidates and their scores and rankings must be publicly displayed online (on the PSB's portal since 2017). The 2012 measures added that "If the application is not accepted the reason will be given" (article 15). From 2017 additional progress in terms of accountability was made: during the publicity period (of 7 days), eligible applicants had to submit materials confirming their fulfilment of the criteria for

[54] Information disclosure makes information on government action publicly available and is meant to improve bureaucratic accountability and reduce corruption (Stromseth, Malesky, and Gueorguiev 2017; Chen and Greitens 2022). Online responsiveness is the use of information and communication technologies to collect public opinions, interact with citizens, and increase public participation (Meng and Yang 2020).

scoring points, and, if ineligible applicants had any objection to the results, they were able to apply to the Bureau for review. They could also file administrative lawsuits against decisions, with good chances of winning the case if there had been breaches of procedure. The automation of decision-making processes, however, shifted potential causes of discontent from the way individual applications were treated towards the rules of the game itself. In the case of the points systems, this meant towards the policy choices made regarding the index table and the weight given to different indicators, as well as to the way the government communicated about the points system. The last section examines citizens' complaints within the framework of Open Government.

Citizens' Requests for Information, and Automated Responses

Within the framework of Open Government, the Shenzhen government tends to consider that dissatisfaction can be reduced by either better explaining the policy or by rectifying gaps and contradictions within the policy after these insufficiencies have been pointed out in public feedback on its designated forums and official channels such as message boards for leaders. This section looks at the complaints that are voiced both on the official platforms and in commercial forums run mostly by private counseling agencies. People rarely contest the rationale of the points systems per se, or the *hukou* system itself (the few criticisms of the system are issued by NGOs and Chinese scholars); rather, most postings, especially on official channels, are framed as suggestions for improving the points systems (as also noted by Brown 2021). Close attention to the content, however, shows that online comments and requests for information do have a critical dimension. The complaints question the government's claim to reasonable and open governance, since the most frequent comments relate to a lack of clear information regarding thresholds and quotas. Moreover, the shutting down of the system in Shenzhen in 2020 further fueled uncertainty and anger arising from the absence of information. Framing of the points systems in terms of scientific fairness, however, allows any substantial critique to be smothered.[55]

Online forum netizens recurrently raised questions about the workings of the systems, particularly the exact number of points one needed to be within the quota. The quota of yearly admissions was generally announced only once the channel of applications was open, making it impossible to know before applying. In Shenzhen, the very existence of a quota seems to have been ignored by many until recently.

[55] Many posts criticize the favoring of the rich and educated; I discuss this class aspect elsewhere (Trémon 2025).

Early policy texts did not use the term quota (*edu*); instead, they alluded to it, stating that *hukou* would be granted to those who had an eligible score according to "*the Shenzhen Municipal Instrument of Skilled Workers Recruitment*" (Shenzhen 2010) and "*in accordance with the procedures*" (Shenzhen 2012). The 2016 Regulations mentioned the quota limitation for the two talent channels, and for "family relocation", which had not been subject to a quota until then (Shenzhen 2016, articles 5, 10 and 11). Only in 2017 did the term feature explicitly at the beginning of the new trial regulations which enumerated a long list of conditions, concluding with the criterion: "and approved for household entry within the annual plan's quota" (Shenzhen 2017b, article 2).

Even if the existence of a quota was now explicitly acknowledged, there was very little information available on how many points applicants in actual fact needed to accrue to have a chance of being accepted for Shenzhen *hukou*. This was due to the discrepancy between the eligibility threshold, which was set at 100 points in 2012, and the actual total score needed for approval within the quota. Applicants meeting the basic threshold were ranked according to their total number of points, and only the 10,000 applicants ranking highest were accepted for *hukou* registration. Online counselors tended to give advice based on the previous year's quota and the real threshold. The *Hukou* Change Observation Blog, for instance, which issued a handy guide explaining the policy by taking members of the Peppa Pig family as illustrations for different cases, revealed that "the score for the 10,000th person in 2017 was 307 points, which is equivalent to about 15 years of rent and social security". In 2018, the number of quotas for household registration was still 10,000; therefore, if at September 2018 Mama Pig had been renting a house in Shenzhen for 15 years and 1 month, and had paid pension insurance for 15 years and a month, she had a good chance of seeing her *hukou* application accepted (*Hukou* Change Observation Blog 2018).

This was a source of frustration for those nominally (on paper) eligible but actually ineligible. It was probably the reason the basic threshold was removed and emphasis placed on "competition for the best" (*jingzheng zeyou*) (Shenzhen 2021). The minimal length of time in residence and payments into social security increased to ten years instead of five. These higher thresholds made the system seem less accessible but were more consistent with the reality of the system's actual functioning. Nevertheless, uncertainty remained about whether the quota would change, since it varied on a yearly basis along with annual plans. In 2020 and 2021, prior to the announcement of the new measures, wild speculation circulated concerning the possible rise in the threshold and decrease of the quota. Lack of knowledge and uncertainty about the actual number of points needed to be accepted into a Shenzhen household fed online discussions encouraging people to increase their number of points by all and any means.

This widespread uncertainty was compounded by the shutting down of the system at the beginning of 2020. The Public Security Bureau issued a notice on

January 23, 2020 stating that, starting from 17:00 on February 1, 2020, Shenzhen would suspend the points entry channels. This was the date when the lockdown started in the city of Wuhan where the Covid pandemic had broken out. This coincidence of dates led many people to assume that the suspension was due to Covid, but this was not the case. The reason was that the trial measures of 2017 had come to an end, having been valid for three years in response to the NUP setting 2020 as the limit for reaching the plan's goal. People had to wait until June 2021 to get information about the direction of the new points policy (the introduction of talents upon simple approval went on as usual). It seems the municipal authorities were awaiting an announcement from the new NUP to revise the policy accordingly. The Plan was released on June 22, and the Municipal Party Committee of Shenzhen released its suggestions for the city's fourteenth five-year Plan in October 2021.

Among the many posts published from May 2020 to September 2022 on the Shenzhen message board for leaders, 28 messages dealt with the points systems for accessing *hukou* or public-school places (all except one written by non-*hukou* holders). The critiques, which were often quite vehement in tone, mostly related to the closure of the points-based channel. In 2020, they criticized not so much the closure itself but rather the lack of information about the timing of the reopening. As time went by, they increasingly demanded the reopening of the system, or that clear reasons be given for its closure, criticizing the opacity and the secrecy. They based their critiques on the very principles that points systems are supposed to reflect: the principles of scientific fairness and transparency that the city government espoused to justify them.

Although most messages were specifically related to the immediate context of the suspension, netizens reported on their individual situations, explaining the impact of the interruption on their lives, including their emotional distress. Some noted that they had worked hard to earn a part-time degree that would increase their score and their chance of getting *hukou*, while others simply mentioned that they had been working hard for many years to get their application ready and were now left in limbo. Expressing feelings of disappointment and anger, and of anxiety and despair, they emphasized their inability to plan for their futures and that of their children, children's futures being the most common reason given for their hopes of access to urban citizenship. By insisting on this, they pointed out the contradiction between the intended outcomes of the policy, that is, granting *hukou* to those who led stable lives in Shenzhen, and its abrupt suspension. They also made ironic remarks about inconsistencies in the policy, emphasizing the discrepancy between the policy's principles of openness and city slogans such as "Shenzhen speed" (urging the city government to respond rapidly).[56]

[56] See note 2.

They received a standardized response explaining that “the system is currently under revision”, or reasserting some of the policy’s principles. Before announcing the new trial measures in the fall of 2021, however, Shenzhen authorities first issued a draft on which they solicited opinions from the public. Opinions could be sent by electronic or postal mail by June 23. It took the municipal authorities only one day to examine them, with their response posted on June 25 (Shenzhen MHRSSB 2021). On online forums, most netizens showed no interest in sending their opinions to the government; what they wanted was certainty (WeChat 2021a). In reaction to the announced tightening of the *hukou* policy, most people asked for more timely information, no matter the content, in order to be able to adjust their life plans. Some asked that a transition period be implemented in order to make up for the shutting down of the system. One netizen announced that he would be sending an opinion to the government asking for “buffer time”; otherwise, he argued, the policy would be “unreasonable” (WeChat 2021b). The prolonged absence of information regarding the shutdown followed by an announcement of policy tightening was met with much despair.

The only available online data showing the feedback received by the Shenzhen government is a table that summarizes the propositions, states whether they are accepted or not, and gives a short explanation. One potentially subversive demand was that people meeting the basic threshold be granted *hukou* without being ranked. This was rejected based on the principle of “selection of the best”. Most feedback related to very specific issues such as the number of points for various criteria. Among those that implicitly pleaded for a policy relaxation, was a request for increasing the value of social security points. This was not accepted because of the principle that “the weights of various scores” should be “kept in a reasonable proportion”. Another request was to increase the score of part-time undergraduate education. This was declined on the grounds that “the current score conforms with the rationality of the score setting”.

In short, the policy principles determined the irrevocability of the framework, principles that were presented as rational and reasonable, within which adjustments could be made. In the case of schools, all those who had filed lawsuits, arguing that being denied a school place was inconsistent with the right to an education backed by the law on compulsory education, lost their cases because policy documents at all levels of government systematically tied the allocation of resources to each city’s “carrying capacity”. Points systems form a self-referential system, similar to the legal system, in which netizens throw terms used to justify them back at the government, and the government in response invokes the principles it has put forward to justify its actions.

Conclusion

During 2022 anti-zero-Covid online and street demonstrations, Chinese citizens argued that the government policy had no scientific basis, that it was *wu kexue* (BBC 2022). Points systems' claims to "scientificalness" are also under netizen fire. Their very purpose is to achieve the goal of adopting the scientific outlook on development, that is, to ensure the restructuring of large cities' economies reliant on unskilled migrant labor into high tech service economies reliant on skilled and educated "talents". They aim at optimizing the quality of each city's population by matching population size and structure to economic needs and "carrying capacity". Points systems are a tool for achieving this planned development, and are legitimized by the prevailing scientist notion that quantified measures are intrinsically scientific.

The digitalization of points systems has rested on the construction of city-wide databases that have fueled the gradual extension of this method to most categories of *hukou* applicants, and to other domains of application such as public-school places. Although, in Shenzhen, points channels were initially intended for deserving ordinary workers, and were distinct from direct approval channels for talents, they have progressively hybridized, with points channels being ambiguously re-labeled "talents". Even though they differentiate between individuals based on their capital and demographic features such as age, the comparison allowed by the scoring system and resulting unified procedures is described as scientific.

The legitimizing effect of points systems not only rests in their claimed "scientificalness" but also in the notion that a scientific approach ensures fairness. China's tradition of objective evaluation of moral comportment, which was enhanced under Maoism and continues under socialist neoliberalism, plays a great part here. Points systems are a tool for evaluating deserving urban residents, but also for selecting those most capable of participating in the municipal budget and in city life through their monetary contributions and voluntary donations. Because they are expected to incentivize people, points systems have to be explained in a way that is understandable to the public.

This chapter has suggested that the adjective "reasonable" justifies the policies by appealing to common sense but also by shifting the emphasis from science to governance. This is possibly because science is an insufficient source of legitimation. In spite of points systems' purported calculability and accountability, complexity and uncertainty are contained within their very workings. They are generated by the systems' constant readjustments, which are not just due to the experimental nature of the policy, but also to the aims and techniques of population governance.

Paradoxically, the good governance principles encapsulated in "scientific fairness" do allow for the voicing of critique based on these very principles, when netizens complain about the policies' opacity for instance. Yet, points systems' governance functions in a self-referential logic, as in cybernetics: the policy is scientific and reasonable because it uses quantified tools and informational technologies, but the tools for reaching the policy goals must be quantified and informational. The principles are reasonable because they are fair and based on the calculation of available resources, and vice versa. This thoroughly depoliticized circularity precludes critique of such governance practices' operation and rationale.

Thus far scientific fairness has not become a master concept in CCP thought, and it will probably never be considering the way Xi Jinping's actions and thought have overshadowed Hu Jintao's concepts, although scientific development remains a goal. Despite this, the principles of "scientific and reasonable" allocation of resources through points systems that have shaped population management policies over the past dozen years are likely, considering the continued importance of planning, and the extensive information infrastructure required to build China's population policies.

References

Bakken, Børge. 2000. *The Exemplary Society: Human Improvement, Social Control, and the Dangers of Modernity in China*. Oxford, New York: Oxford University Press.

Barnett, Clive, Nick Clarke, Paul Cloke, and Alice Malpass. 2008. "The Elusive Subjects of Neo-Liberalism: Beyond the Analytics of Governmentality." *Cultural Studies* 22 (5): 624–53.

BBC 2022. "Shanghai outbreak causes China's epidemic prevention policy to waver" 11 April, in Chinese: https://www.bbc.com/zhongwen/simp/chinese-news-61065997

Brown, Junius F. 2021. "Development and Citizenship in the Chinese 'Mayor's Mailbox' System." *Asian Survey* 61 (3): 443–72.

Chen, Jidong, Jennifer Pan, and Yiqing Xu. 2016. "Sources of Authoritarian Responsiveness: A Field Experiment in China." *American Journal of Political Science* 60 (2): 383–400.

Chen, Tao, Zhehao Liang, Hongtao Yi, and Si Chen. 2023. "Responsive E-Government in China: A Way of Gaining Public Support." *Government Information Quarterly* 40 (3): 101809.

China Daily 2010. "Scientific Outlook on Development", 8 September, xinhua/chinadaily.com.cn/people.com.cn

China.org.cn 2021. "Scientific Outlook on Development", updated 21 September, http://www.china.org.cn/m/english/china_key_words/2021-09/21/content_77765465.html

Creemers, Rogier. 2018. "China's Social Credit System: An Evolving Practice of Control." SSRN Scholarly Paper, https://doi.org/10.2139/ssrn.3175792.

———2020. "The Ideology Behind China's AI Strategy." In *The AI Powered State: China's Approach to Public Sector Innovation*, Hessy Elliott ed., 63–69. Nesta. Accessed July 14, 2023. https://www.nesta.org.uk/documents/1860/Nesta_TheAIPoweredState_2020.pdf.

Daum, Jeremy. 2017. "China through a Glass, Darkly." *China Law Translate* (blog). December 24. https://www.chinalawtranslate.com/china-social-credit-score/.

Desrosières, Alain. 2010. *La politique des grands nombres: Histoire de la raison statistique*. Paris: La Découverte.

Distelhorst, Greg, and Yue Hou. 2017. "Constituency Service under Nondemocratic Rule: Evidence from China." *The Journal of Politics* 79 (3): 1024–40.

Dong, Yiming, and Charlotte Goodburn. 2020. "Residence Permits and Points Systems: New Forms of Educational and Social Stratification in Urban China." *Journal of Contemporary China* 29 (125): 647–66.

Fairclough, Norman. 2013. *Critical Discourse Analysis*. London: Routledge.

Foucault, Michel. 1991. "Governmentality." In *The Foucault Effect: Studies in Governmentality*, Graham Burchell, Colin Gordon, and Peter Miller, eds. 87–104. London: Harvester Wheatsheaf.

Göbel, Christian, and Jie Li. 2021. "From Bulletin Boards to Big Data: The Origins and Evolution of Public Complaint Websites in China." *Journal of Current Chinese Affairs* 50 (1): 39–62.

Goddard, Stacie E., and Ronald R. Krebs. 2015. "Rhetoric, Legitimation, and Grand Strategy." *Security Studies* 24 (1): 5–36.

Greenhalgh, Susan. 2003. "Science, Modernity, and the Making of China's One-Child Policy." *Population and Development Review* 29 (2): 163–96.

Greenhalgh, Susan, and Edwin A. Winckler. 2005. *Governing China's Population: From Leninist to Neoliberal Biopolitics*. Stanford: Stanford University Press.

Greenhalgh, Susan, and Li Zhang, eds. 2020. "Governing Through Science: The Anthropology of Science and Technology in Contemporary China." In *Can Science and Technology Save China?*, 1–24. Ithaca NY: Cornell University Press.

Guangdong Provincial Government. 2010. "Guiding Opinions on Carrying Out the Work of Point-based Systems for Peasant Workers' Entrance into Towns and Cities" No. 32, 5 July 2010 http://www.gd.gov.cn/gkmlpt/content/0/139/post_139081.html#7

Gueorguiev, Dimitar D. 2021. *Retrofitting Leninism: Participation without Democracy in China*. New York: Oxford University Press.

Guo, Zhonghua, and Tuo Liang. 2017. "Differentiating Citizenship in Urban China: A Case Study of Dongguan City." *Citizenship Studies* 21 (7): 773–91.

Hacking, Ian. 1991. "How Should We Do the History of Statistics?" In *The Foucault Effect*, Graham Burchell, Colin Gordon, and Peter Miller, eds, 181–95. Chicago: The University of Chicago Press

Hukou Change Observation Blog. 2018/ "Guide for Entering Shenzhen Household: Six hukou acquisition methods", August 1, http://hugaiguancha.blog.caixin.com/archives/185167

Kipnis, Andrew. 2007. "Neoliberalism Reified: Suzhi Discourse and Tropes of Neoliberalism in the People's Republic of China." *The Journal of the Royal Anthropological Institute* 13 (2): 383–400.

Li, Limei, Si-ming Li, and Yingfang Chen. 2010. "Better City, Better Life, but for Whom?: The Hukou and Resident Card System and the Consequential Citizenship Stratification in Shanghai." *City, Culture and Society* 1 (3): 145–54.

Luhmann, Niklas. 1995. *Social Systems*. Writing Science. Stanford, Calif: Stanford University Press.

Munro, Donald J. 1971. "The Malleability of Man in Chinese Marxism." *The China Quarterly* 48: 609–40.

National Bureau of Statistics 2021 "Basic population situation of megacities and super-large cities in the seventh national population census" September 16, https://www.stats.gov.cn/zt_18555/zthd/lhfw/2022/lh_jjsh/

O'Donnell, Mary Ann, Winnie Won Yin Wong, and Jonathan P. G. Bach, eds. 2017. *Learning from Shenzhen: China's Post-Mao Experiment from Special Zone to Model City*. Chicago: The University of Chicago Press.

O'Neil, Cathy. 2016. *Weapons of Math Destruction: How Big Data Increases Inequality and Threatens Democracy*. First edition. New York: Crown.

Papagianneas, Straton. 2023. "Smart Governance in China's Political-Legal System." *China Law and Society Review* 6 (2): 146–80.

Rodenbiker, Jesse. 2021. "Making Ecology Developmental: China's Environmental Sciences and Green Modernization in Global Context." *Annals of the American Association of Geographers* 111 (7): 1931–48.

Rose, Nikolas. 1991. "Governing by Numbers: Figuring out Democracy." *Accounting, Organizations and Society* 16 (7): 673–92.

Rose, Nikolas, and Peter Miller. 2010. "Political Power beyond the State: Problematics of Government: Political Power beyond the State." *The British Journal of Sociology* 61: 271–303.

Scott, James C. 1998. *Seeing like a State: How Certain Schemes to Improve the Human Condition Have Failed.* New Haven CT, London: Yale University Press.

Shen, Yang, and Bingqin Li. 2020. "Policy Coordination in the Talent War to Achieve Economic Upgrading: The Case of Four Chinese Cities." *Policy Studies* (1): 1–21.

Shenzhen MHRSSB Municipal Human Resources and Security Bureau 2010. Human Resources and Social Security Bureau, Annual Report on Government Information Disclosure in 2009, 4 March 2010 http://hrss.sz.gov.cn/xxgk/xxgknb/content/post_2007852.html

Shenzhen MHRSSB 2021. Explanation on the adoption of the "Shenzhen Approved and Point-based Talent Introduction and Household Implementation Measures (Draft for Comment)" after soliciting opinions, feedback released on 25 June, http://hrss.sz.gov.cn/hdjlpt/yjzj/answer/12376

Shenzhen Municipal Government 2005. Opinions on Strengthening and Improving Population Governance in Shenzhen and Five Supporting Documents, 1 August, http://www.sz.gov.cn/zfgb/2005/gb449/content/post_4977073.html

Shenzhen Municipal Government 2010. Notice on Printing and Distributing the Trial Method for Migrant Workers' Points-Based Access to hukou in Shenzhen. No. 70, 30 August, http://www.sz.gov.cn/zfgb/2010/gb711/content/post_4953653.html

Shenzhen Municipal Government 2011. Notice on Printing Shenzhen's Twelfth Five-Year Plan n°86, July 11, http://www.sz.gov.cn/zfgb/2011/gb762/content/post_4948391.html

Shenzhen Municipal Government 2012. Interim Measures for Point-based Household Registration of Migrant Workers in Shenzhen, http://www.sz.gov.cn/cn/xxgk/zfxxgj/zcfg/szsfg/content/post_6582096.html#

Shenzhen Municipal Government 2016. Notice on Shenzhen Household Registration Relocation Regulations, n° 59, 23 August, http://www.sz.gov.cn/zfgb/2016/gb968/content/post_4963208.html

Shenzhen Municipal Government 2017a. Trial regulations, Interim Measures for Point-based *hukou* Acquisition in Shenzhen, n°1 13 July, http://fgw.sz.gov.cn/attachment/0/297/297321/4591139.pdf

Shenzhen Municipal Government 2017b. Notice of the on Printing and Distributing the 'Thirteenth Five-Year Plan' for the Development of Population and Social Undertakings, 11 January, http://www.sz.gov.cn/zwgk/zfxxgk/zfwj/szfwj/content/post_6577335.html. *The Plan itself was issued on November 9.*

Shenzhen Municipal Government 2021. Explanation on the Adoption of the Several Regulations on the Relocation of Household Registration in Shenzhen (Draft for Comment), 18 June 2021.

Shenzhen Municipal Public Security Bureau 2023. Notice on the Issuance of the "Shenzhen Points-based Household Registration Method" June 25, https://ga.sz.gov.cn/gkmlpt/content/10/10666/post_10666484.html#25119

Shenzhen Municipal Market Supervision Administration 2023. Shenzhen Special Economic Zone Social Credit Regulations, February 28, https://amr.sz.gov.cn/xxgk/zcwj/scjgfg/xyjg/content/post_10450868.html

Shore, Cris, and Susan Wright. 2015. "Audit Culture Revisited: Rankings, Ratings, and the Reassembling of Society." *Current Anthropology* 56 (3): 421–44.

Sigley, Gary. 2006. "Chinese Governmentalities: Government, Governance and the Socialist Market Economy." *Economy and Society*, 35 (4): 487–508.

Sklair, Leslie. 1991. "Problems of Socialist Development: The Significance of Shenzhen Special Economic Zone for China's Open Door Development Strategy." *International Journal of Urban and Regional Research* 15 (2): 197–215.

Smart, Alan, and Josephine Smart. 2001. "Local Citizenship: Welfare Reform Urban/Rural Status, and Exclusion in China." *Environment and Planning A: Economy and Space* 33 (10): 1853–1869.

Solinger, Dorothy J. 1999. *Contesting Citizenship in Urban China: Peasant Migrants, the State, and the Logic of the Market*. Studies of the East Asian Institute, Columbia University. Berkeley: University of California Press.

State Council 2006. Opinions About Solving the Issues of Peasant Migrant Workers, No 5, 31 January, http://www.gov.cn/gongbao/content/2006/content_244909.htm.

State Council 2007. People's Republic of China Regulations on Information Disclosure, promulgated on April 5, https://www.gov.cn/zwgk/2007-04/24/content_592937.htm

State Council 2012. Notice on Actively and Steadily Promoting the Reform of the Household Registration System, 23 February, http://www.gov.cn/zhengce/content/2012-02/23/content_1097.htm

State Council 2014a. National New Urbanization Plan (2014-2020), 16 March 2014, http://www.gov.cn/zhengce/2014-03/16/content_2640075.htm

State Council 2014b. Opinions on Further Promoting the Reform of the Household Registration System, No. 25, 24 July 2014, http://www.gov.cn/zhengce/content/2014-07/30/content_8944.htm"

Strathern, Marilyn, ed. 2000. *Audit Cultures: Anthropological Studies in Accountability, Ethics, and the Academy*. European Association of Social Anthropologists. London New York: Routledge.

Stromseth, Jonathan, Edmund Malesky, and Dimitar D. Gueorguiev. 2017. *China's Governance Puzzle: Enabling Transparency and Participation in a Single-Party State*. Cambridge UK, New York: Cambridge University Press.

Supiot, Alain. 2015. *La Gouvernance Par Les Nombres: Cours Au Collège de France, 2012–2014*. Paris: Fayard.

Thornton, Patricia 2011. "Retrofitting the Steel Frame: From Mobilizing the Masses to Surveying the Public" in Sebastian Heilmann and Elizabeth J. Perry, eds. *Mao's Invisible Hand, The Political Foundations of Adaptive Governance in China*, Harvard University Press, 237–258.

Trémon, Anne-Christine. 2022. "Retour à la frugalité paysanne? Développement urbain et civilisation écologique, une réflexion à partir du cas de Shenzhen." *Monde chinois* 68 (1): 97–121.

Trémon, Anne-Christine. 2025, forthcoming. "Spatial Governmentality and the Rhetoric of Scarce Resources: 'Scientific and Reasonable' Points Systems in Shenzhen (China)" *Environment & Planning A: Economy & Space.*

Trémon, Anne-Christine, Cliff Chen, and Liu Na. 2024. Citizenization or Talent Recruitment? The Development of China's Points-Based Hukou Channels, *Journal of Contemporary China*, early online publication. DOI: 10.1080/10670564.2024.2360058

Unger, Jonathan. 1984. "Remuneration, Ideology, and Personal Interests in a Chinese Village, 1960–1980." *International Journal of Sociology* 14 (4): 3–26.

Wan, Yi, and Edward Vickers. 2022. "Towards Meritocratic Apartheid? Points Systems and Migrant Access to China's Urban Public Schools." *The China Quarterly* 249 (March): 210–38.

Wang, Fenglong, and Yungang Liu. 2018. "Interpreting Chinese *Hukou* System from a Foucauldian Perspective." *Urban Policy and Research* 36 (2): 153–67.

Wang, Xiang. 2020. "Permits, Points, and Permanent Household Registration: Recalibrating *Hukou* Policy under 'Top-Level Design.'" *Journal of Current Chinese Affairs* 49 (3): 269–90.

Watts, Galen. 2022. "Are You a Neoliberal Subject? On the Uses and Abuses of a Concept." *European Journal of Social Theory* 25 (3): 458–76.

WeChat 2021a. "Find out now! The latest news of Shenzhen point-based household registration is here! So many changes," 3 June, anonymous posting, not dated. https://mp.weixin.qq.com/s/_Y7XCKe1MG-M5KxaZGjvOkQ

WeChat 2021b. "Heavy News, Shenzhen intends to tighten the settlement policy!" 26 May, anonymous posting, not dated, https://mp.weixin.qq.com/s/saSPtJCkYpqRoYf3MpzLZA

Wong, Linda, and Huen Wai-Po. 1998. "Reforming the Household Registration System: A Preliminary Glimpse of the Blue Chop Household Registration System in Shanghai and Shenzhen." *International Migration Review* 32 (4): 974.

Woodman, Sophia, and Zhonghua Guo. 2017. "Introduction: Practicing Citizenship in Contemporary China." *Citizenship Studies* 21 (7): 737–54.

Zhang, Chenchen. 2018. "Governing Neoliberal Authoritarian Citizenship: Theorizing *Hukou* and the Changing Mobility Regime in China." *Citizenship Studies* 22 (8): 855–81.

———. 2020. "Governing (through) Trustworthiness: Technologies of Power and Subjectification in China's Social Credit System." *Critical Asian Studies* 52 (4): 565–88.

Zhang, Li. 2012. "Economic Migration and Urban Citizenship in China: The Role of Points Systems." *Population and Development Review* 38 (3): 503–33.

Zhang, Li, and Meng Li. 2016. "Local Fiscal Capability and Liberalization of Urban *Hukou*." *Journal of Contemporary China* 25 (102): 893–907.

Zuboff, Shoshana. 2019. *The Age of Surveillance Capitalism: The Fight for a Human Future at the New Frontier of Power*. New York: Public Affairs.

CHAPTER 8

Queer Social Sorting: Control and Criminalization in China's LGBTQIA+ Activism

Ausma Bernot

Abstract

On the surface, China's authorities maintain formal decorum and formulaic support for LGBTQIA+ communities. Non-discrimination can be argued on the basis of the thawing of national laws that restrict LGBTQIA+ communities, such as the recent 2022 regulations lowering the minimum age requirements for gender-affirming surgery from twenty to eighteen. At the same time, suppression of LGBTQIA+ activism persists under this thin layer of political decorum. The state-sponsored control of queer activism is now increasingly linked to two elements: China's call to return to traditional gender roles and *social sorting* of queer activism. Queer social sorting is achieved through the interconnected tools of legal and regulatory frameworks, public and state security monitoring and harassment, and digital surveillance.

Keywords: Queer social sorting; Queer surveillance; LGBTQIA+; Activism

Introduction

This chapter examines state-facilitated surveillance and control practices that have been deployed in the People's Republic of China (PRC) to control LGBTQIA+ activist and advocate groups since 2013. These practices vary from legal and regulatory to political–ideological and technological. Alongside physical surveillance and policing practices, automated decision-making is now increasingly deployed to ensure that control over queer communities is effectively implemented. This chapter reflects on how China's experimentation with automation technologies of datafication and algorithmic governance entrenches patriarchal and normative gender roles in the ordering of social life in China.

As this chapter will show, queer community organizing exists in a dubious space between surface-level state commitment to non-discrimination and on-the-ground realities that constrict LGBTQIA+ organizing. The Party-state has maintained plausible deniability and political decorum by avoiding the direct prohibition of LGBTQIA+ groups and their activities and instead ostracizing, controlling, and even criminalizing essential activities required to sustain LGBTQIA+ groups (fundraising, for example). Such practices must be identified and challenged at local and global levels.

Before delving into the research findings, it is important to acknowledge that LGBTQIA+ activism in China does not encompass all queer life in the country. In reviewing three decades of Chinese LGBTQIA+ activism, Stephanie Yingyi Wang (2021) argued that the NGO-ization of queer activism only emerged in the 1990s "as a distinct cultural, social, and political product in a modernizing China with increasing transnational connections in the wake of market reforms" (91). Wang describes the domestic development of LGBTQIA+ communities mixed with transnational queer politics. Indeed, the participation of international non-government organizations (NGOs) has been important in obtaining funding (Hildebrandt 2012); however, such funding is closely tied to a vision of rights-based activism and advocacy, with a distinct focus on "a progressive reform agenda" (Kellogg 2012). State actors now label such activism efforts as influenced by "hostile foreign forces" (Zhao 2022).

In this chapter, "queer" and "LGBTQIA+" are used interchangeably. While these two terms are commonly used by queer communities in the PRC, there is a long history of queer people creating and using indigenous terms of self-identification, such as the subverted term "comrade" or *tongzhi* (同志), a socialist word-turned-queer (Bao 2011). *Tongzhi* was mandated as a form of address by Mao Zedong in the 1950s. It was first used by the queer community at the Gay and Lesbian Film Festival in Hong Kong in 1989, after which it spread in popularity. Many other terms, both historical (e.g. Kang 2009), currently used (such as *lala* 拉拉), and creatively inducted (for example, kua'er 跨儿), adorn the modern queer dictionary.

The findings of this chapter build on twenty-six interviews, conducted from 2021 to 2022, with LGBTQIA+ community organizers and activists from 12 provinces in mainland China. Recruitment efforts consciously focused on ensuring that trans and gender-diverse populations were fairly represented, these communities being the most marginalized in terms of legal protections, access to education, and social and medical services (Bernotaite et al. 2017). Ten of twenty-six interviewees identified as trans or genderqueer. To protect the identities of the interviewees, I use pseudonyms when quoting parts of our conversations. The data is de-identified and I have redacted the specific mentions of organizations and names of provinces/cities that have one primary LGBTQIA+ organization. Such strict redaction measures ensure that the actions taken by public and state security actors can be reported in detail while preserving interviewee confidentiality. Where the analysis found meaningful regional differences, those are specifically noted in-text without a direct mention of province or city names.

Despite the limited sample size, the interviews reached sensitive communities spread across several regions, providing rare empirical evidence from China at a time when crucial research is severely restricted and LGBTQIA+ organizations are closing *en masse.* In a brief survey of the organizational closures, several painful events stand out: the May 2023 closure of the Beijing LGBT Centre (Wang 2023), the 2021 closure of LGBT Rights Advocacy China, the wave of WeChat censorship of university-based queer groups that caused disruptions to their organizations (Bernot 2022; Wilhelm 2022), and the abrupt 2020 closure of Shanghai Pride due to "security concerns" (Wilhelm 2022). Accompanying the closures, the organizations' statements were written in "queer double-speak" (Bernot 2023), alluding to pressures, tensions, giving-up, and losing control.

This chapter delves into the surveillance and monitoring activity that precedes the closing down of organizations and the shrinking of communities, facilitated by state and public security organs. I argue that, since 2013, the suppression of LGBTQIA+ groups and organizations has become institutionalized, and criminalization has followed. The institutionalization has followed a dual strategy: implementation of an agenda of broad gender-setting by the Party-state, and use of a suite of covert surveillance strategies, or *queer social sorting* techniques (Bernot and Davies 2023). Increasingly, automatic decision-making accompanies regulatory and legal instruments to ensure that suppression occurs on the ground. Several important parameters that trigger suppression can be discerned from my chapter's focus, namely, international funding, non-registered group status, advocacy-leaning organizational profile (as opposed to community services-leaning), high online visibility, and activities during politically-sensitive times. The chapter first provides the context of "moralized" gender roles and then discusses queer social sorting. The last section of the chapter provides examples of automation in the surveillance and control of LGBTQIA+ groups.

Locating LGBTQIA+ Groups within Political–Ideological Surveillance Practices in China

LGBTQIA+ people do not fit the modern political–ideological gender imagery reflected in "the China dream" (中国梦). Xinhua News (2021) published a statement from Xi Jinping himself stating that the role of women in the exercise of nation-building was to strike "a balance between family and work in order to become women of the new era who can take up social responsibilities while contributing to their families." The categorization of queer social groups in the larger framework of Chinese practices involves classifying and contextualizing populations according to the Party-state's imaginary of a Chinese identity. This encompasses two key aspects: first, the utilization of an "othering" strategy, which manifests as internal differentiation within the Chinese populace (within the scope of Chinese nationalism and the Chinese dream; 中国梦), and operates as a geopolitical strategy (framing queer populations as national security threats). The second aspect is the implementation of surveillance techniques involving monitoring, regulatory measures, and repressive tactics, including criminalization and monitoring of LGBTQIA+ groups.

Across generations, there is a growing acceptance of more liberal-egalitarian gender attitudes, with the trajectory of change significantly dependent on urban, rural, and migrant identities (Yang 2023). The formal state narrative, however, introduces tension to those increasingly strong gender equity beliefs, artificially framing the "gender others" such as feminist and LGBTQIA+ groups as anti-China. The activities that LGBTQIA+ organizations generally undertake (such as fundraising and social awareness campaigns) are now criminalized, justifying the need for state-facilitated surveillance.

Perhaps one of the most significant turning points in establishing "moralized" gender rules was the 2020 census. The census confirmed that, despite the one-child policy being relaxed to a two-child policy in 2015, China's population grew at its slowest rate since the 1950s (Ren 2022; Yang 2023). As the birth rates in China continue to drop despite the elimination of the one-child policy, the Party-state is (re)prioritizing the political–ideological framing of the nuclear family unit as the bedrock of society. Despite the persistent and rising child penalty— that is, economic and career-related disadvantages that parents, particularly mothers, face after having children (Zhou et al. 2022)—most state-backed efforts are directed at hastening early marriages and making it more difficult to get a divorce (Ren 2022). As Rujun Yang (2023:1) argues, egalitarian and essentialist beliefs did not coincide with China's economic development; they did not reflect "the resilience of male-primacy ideology."

Child planning policies have resulted in indirectly "othering" queer populations, putting LGBTQIA+ people out of favor with state efforts to "make more babies for the GDP" (*Xiaoyue, lesbian activist*). As one interviewee bluntly asserted in their interview:

> Now, [the government of China] wants people to have three children, and they think that LGBT+ people are not good people because they make everyone gay. (*Wang Tie, genderqueer activist*)

Others mirrored that opinion:

> From the official media, there's this general trend that there is increasing promotion of this idea that families are still very important and having kids is important. ... Women should get married. They should have kids. (*Zhao Yun, queer woman activist*)

The monitoring, control, and suppression strategies directed at LGBTQIA+ activism in China were first applied to feminist activism, most notably in the second half of the 2010s. Personal reports of volunteering in queer feminist activism revealed that something shifted in 2015 with the arrest of the Feminist Five—five women who were planning to hand out stickers opposing sexual harassment on public transport on International Women's Day on March 8. In feminist and queer activist circles, there was suddenly more fear, more awareness of surveillance, and rushed migration to encrypted platforms. In the years post 2015, activists were noting the first upsetting signs of the growing surveillance spotlight: public security interrogations, whispers of public security harassment of landlords to issue eviction notices, and new laws making it harder to organize—or to do so legally. Clear signs of severe censorship of feminist content on social media became apparent as ultra-nationalist misogynist voices were permitted to bully and drown out the content that remained online (Lindberg 2021; Yin and Sun 2021). While some scholarship has found evidence of uneven, "ambivalent" governance of censorship (Gu and Heemsbergen 2023), as well as evidence of smart counter-censorship strategies deployed by activists (Yang 2021), the online space remains difficult for Chinese feminists.

Similar offline and online monitoring, surveillance, and suppression strategies were then adopted to control LGBTQIA+ activist communities. LGBTQIA+ groups were portrayed as an undesired "other" inside the narrative of returning to traditional gender roles, unhelpful at best, and damaging at worst to the cultivation of Chinese society. As Stephanie Yingyi Wang eloquently argues:

> [W]ith the revival of national heteronormative and paternal discourses about the imperative to build the Chinese Dream (中国梦) and enhance Modernity with Chinese Characteristics (中国式现代化), gay male sexualities and homosociality have once again been stigmatized as deviant and responsible for the "boy crisis" (男孩危机)—that is, the corruption of hegemonic masculinity with effeminate gender expressions that threaten the dominant heterosexual nuclear family form—and therefore are a national crisis. (Wang 2023)

The gendered lens that reflected the "otherness" of LGBTQIA+ activists was visible through the interview data:

> [The official government] positioning of people is very...labeled. If you [publicly] show signs of being a feminist, then you're classified as being in the enemy category. And if you have [publicly] expressed that you are homosexual, then you are also classified as the enemy. And if you are transgender, you are also the enemy. So, they put these things that are related to gender... Anything that doesn't fit the norm is considered bad. (*Mingming, pansexual woman activist*)

"Moralized" gender rules are not a new strategy for framing the narrative around queer people in China. History shows that "sex offences" were strictly governed during the Qing dynasty. Such "offences" included marital transgressions (pre- and extra-marital sexual relations), prostitution, and homosexual relations (Ruskola 1993), with a focus on exclusion directed at homosexual cisgender men. Criminalized in 1979, the crime of hooliganism (流氓罪) was an obvious tool for regulating sexuality, with punishments varying in severity from arrests to re-education through labor (Worth et al. 2018). More recently, in writing about the bio-politics of HIV/AIDS medicalization and carving out a space within medical interventions specifically for MSM (men who have sex with men) communities, Haiqing Yu (2016) observed that such a strategy essentializes the HIV-positive as "other," "unruly," and "distanced."

Since the Xi Jinping administration took over, both feminist and LGBTQIA+ movements have often been labeled as influence operations by "hostile foreign forces" (Zhao 2022). In some more extreme examples, LGBTQIA+ people have been linked to state security concerns and foreign espionage. The large state news site People's Daily (2022) writes that the "Five Eyes Alliance" (an intelligence-sharing alliance between the United States, Canada, the United Kingdom, Australia, and New Zealand) committed "deadly crimes" by providing "funding to pro-American individuals and groups under the banner of supporting women's rights, press freedom, and human rights activities." The action was allegedly used as a cover for promoting anti-government ideas and instigating revolutions across Africa and the Middle East. Another news outlet similarly argued that "LGBT is a means used by the United States to export [western] ideology and a tool to serve the globalization of American capital," calling on the people of China to "stick to the road of socialist development with Chinese characteristics, defend traditional marriage, curb the spread of all sexual incest ideologies, including LGBT in our country" to "support the great rejuvenation of the Chinese nation" (Kunlun Ce Research Institute 2023). This is a new strategy that further helps frame LGBTQIA+ groups as a distant "other," even a tool in the hands of hostile foreign governments instigating national threats.

State framing of feminist and LGBTQIA+ populations creates a narrative within which surveillance and investigations against these populations are permitted because they are cast as efforts to support public and state security. Additionally, digital tools, including social media, support public security agencies to collect information about activists and their whereabouts. The next section of this chapter discusses the social sorting aspect of LGBTQIA+ group surveillance.

Queer Social Sorting

The concept of *social sorting* is a theoretical notion from surveillance studies coined by David Lyon in 2002. In analyzing the social impacts of information collection practices, Lyon (2002) found that information databases resulting from surveillance could determine who should be targeted for special treatment, suspicion, eligibility, inclusion, and access. In this way, the databases "sort" the social world in inequitable ways. Lyon's theory is significant as it showcases how covert practices of social sorting can result in isolation, segregation, and marginalization. In theorizing this surveillance power, Lyon's scholarly contribution drew attention to the obfuscation and ambiguity of data-based surveillance practices in the social world. Building on Lyon's idea, I collaborated with Sara E. Davies (2023) to expand the concept of social sorting to queer surveillance in China. This section first considers the collection of data now routinely used for social sorting, and then discusses the dynamics of queer social sorting in China.

The data are first collected via covert surveillance that can eventually become overt, especially in cases where state security actors become involved. Covert surveillance is routine, carried out by both public security actors and by others to whom monitoring and censorship processes are outsourced. The findings of this research suggest that public security actors usually observe the activities of LGBTQIA+ groups via social media, then follow up via phone calls and meetings with LGBTQIA+ activists, a practice colloquially referred to as "drinking tea" or "being invited to drink tea" (喝茶/被喝茶). One of the interviewees who used to work for a national telecommunications provider reported that a part of their job was to routinely transfer a copy of all data to public security (Suisui, *trans activist*). Telecommunications data includes geolocation, unencrypted communications records, and logs of all calls, among other data. In cases where activities that are deemed illegal are registered, the state security stakeholders get involved, and then data collection practices can become significantly more overt and coercive. Specifically, interviewees reported receiving investigatory phone calls to confirm their residential location, their lease arrangements, banking records, and receipts of funding, or reported having their phones and laptops confiscated and forensically analyzed. Activist networks are identified for further investigation through these practices. This chapter contributes to our growing understanding of how the data for queer social sorting are assembled.

To explain how data turns into data-sorting, we have separated the queer social sorting practices into *indirect* and *direct* (Bernot and Davies 2023). *Indirect queer social sorting* practices include direct surveillance via a suite of laws and regulations that do not directly target LGBTQIA+ activism but do have a social impact. Numerous legal and regulatory practices, for example, may not directly target LGBTQIA+ groups, but they have a chilling effect on those groups by making their fundraising practices illegal, while criminalizing their access to information and their international and national partnerships. We argue that data-based surveillance practices are supported by this suite of hastily-published laws and regulations, implying that surveillance practices can fall within the confines of legal practice, and therefore within China's rule of law. In China, LGBTQIA+ group surveillance, control, and suppression are often intentionally obscured through such laws, which foreground and justify data-based social sorting.

The NGO and fundraising management laws are clear examples of indirect social sorting. The laws have created a legal environment in which queer groups cannot legally register and run NGOs; to register an NGO, LGBTQIA+ groups must lodge an official application with the government, approval of which is near impossible due to unrecorded in-person rejections. If the group is unable to register as an NGO, they can register as a commercial enterprise, which allows no tax breaks (Wang 2021). Being formally registered attracts more bureaucratic scrutiny and more rigid government reporting, resulting in "less money ... [and] a greater bureaucratic cost" (Ren and Gui 2022). Additionally, the 2017 Foreign NGO Management Law restricted international NGOs from working in the areas of economics, education, science, culture, health, sport, environmental protection, poverty, and disaster relief. This restriction blocked one of the few remaining pathways for China-based LGBTQIA+ NGOs to obtain funding for their activities (Wang 2021) and foreshadowed LGBTQIA+ group surveillance, examples of which are discussed in detail in the next section.

Direct queer social sorting actively "sorts" LGBTQIA+ groups into unwanted categories via more targeted institutionalized practices and algorithmic censorship. An example of the institutionalized practices is the "sissy ban"—a Notice published in 2021 by the State Administration of Radio and Television "to resolutely reject abnormal aesthetics such as "girly men," to stop catering to "the vulgar and low-brow, to continuously put forward exception works and satisfy the public's spiritual and cultural needs" (The State Radio and Television Administration 2021).

Algorithmic censorship, one of the most common strategies of direct social sorting, is seen via social media that filters much LGBTQIA+ content away from public view (discussed in detail later in the chapter). Social media companies comply with government censorship lists to proactively monitor and censor unwanted LGBTQIA+ content online, sometimes adopting more covert practices, such as shadow banning (Shen 2023) and astroturfing (Miller 2018). Social media censorship is also facilitated by participatory surveillance (Qiao 2022), that is, by ultra-nationalist anti-LGBTQIA+ actors, in response to the Party-state's messaging for a return to normative gender roles. Unrestricted by cyberspace regulations, they now take an active role in reporting and smearing feminist and LGBTQIA+ content.

"Techno-nationalist shaping" facilitates such censorship strategies via media regulations and China's cyber-sovereignty planning (Plantin and de Seta 2019).

Perhaps the most salient of digital space control tools is the 2016 Cybersecurity Law. In 2016, the Central Leading Group for Cyberspace Affairs, chaired by Xi, passed a Law obliging all internet communication platforms to obtain government licenses for operation. Among a suite of requirements, social media platforms are now legally required to monitor real-time content and ensure real-name registration (Creemers 2015). Content deemed unfavorable by state authorities must be censored and reported. This censorship is made more severe because China's internet functions behind the Great Firewall of China. While virtual private networks (VPNs) are a common way to circumvent the Firewall and access encrypted communications platforms, in 2017 Chinese authorities launched a serious crackdown on VPN providers such that it has since been more difficult to install and use VPNs (Qiang 2019). These changes in censorship created a legal and regulatory infrastructure that uses digital spaces to monitor civil society more broadly and LGBTQIA+ communities specifically.

WeChat and Weibo—the two most popular social media companies—are compliant with government-mandated surveillance. WeChat, for example, applies algorithmic surveillance to monitor real-time sensitive content, such as anti-state sentiment and other topics on censorship lists. The content of "sensitive" lists can quickly change, and extreme censorship is applied during critical political periods, such as national government meetings or other politically-sensitive events like COVID-19 outbreaks (Kenyon 2020). LGBTQIA+ communities are specifically targeted on WeChat and Weibo. In June 2021, WeChat, also known as the "everything app," targeted LGBTQIA+ groups in universities and colleges across China, censoring their official accounts on the app (Bernot 2021). In 2018, Weibo announced it would ban homosexual content on the platform to make the site more "harmonious" (Chappell 2018). After citizen pushback, Weibo's stock value dropped, and the platform backtracked from a blanket ban, but it continues to restrict LGBTQIA+ content on the platform (Chia 2019).

The concept of "queer social sorting" explains the dynamics of state-sponsored criminalization of queer groups by the government in power. The prolonged ideological-legal framing of LGBTQIA+ people as "immoral" and "unpatriotic" has stifled the LGBTQIA+ community during Xi Jinping's presidency. Although LGBTQIA+ communities are not officially considered illegal, state suppression efforts of queer communities point towards the covert criminalization of LGBTQIA+ people (Jeffreys 2017), expanding the concept of crime in China. Practices of queer social sorting give substance to political anti-feminist and anti-LGBTQIA+ narratives. Automated decision-making, now occupying an increasingly important—and institutionalized—role in queer social sorting, is discussed in detail in the next section.

Automated Decision-Making to Control the Queer "Other"

The previous sections discussed the context of the Party-state's framing of LGBTQIA+ groups and the strategies deployed for the queer social sorting of LGBTQIA+ activists. This section explores the strategies for controlling queer activism. A particular focus is placed on exploring the role of automated decision-making. Increasingly, automated decision-making is not only supporting the censorship and monitoring of LGBTQIA+ groups but also facilitating the covert criminalization of those groups. The findings presented in this section indicate that the treatment of LGBTQIA+ groups is moving past covert and gradual "illegalization" (Wang 2021) towards institutionalized criminalization, especially when foreign individuals or organizations are involved.

Social media

Social media is a place of contradiction for LGBTQIA+ groups in China, providing digital space for connection and creativity, but also for censorship and restrictions. Social media companies must comply with censorship requirements by employing censors and automated censorship/moderation tools. The interviewees shared stories about the impact of censorship waves on their organizations. This section first discusses the context of social media censorship before considering the impacts on LGBTQIA+ organizing as reported by the interviewees.

Automated social media censorship is not new; what *is* new is the automation of censorship that is now outsourced to private companies by internet laws and regulations; for example, Chinese tech firms are competing to develop automated policing systems that use algorithms to detect suspect patterns in social media activity (Nardi 2019). The previous section explained the complex governance system that now outsources internet censorship tasks to private social media companies via new cyberspace regulations. Those governance changes have had a direct impact on how censorship via social media is implemented. Social media—as well as the governance of social media censorship—is highly decentralized. In their research, King et al. (2013) found that individual social media companies each employ up to 1,000 censors, in addition to 20,000 to 50,000 Internet police (网警) and Internet monitors (网管办), and 250,000 to 300,000 Internet commentators (五毛党) hired to spread positive Chinese Communist Party (CCP) propaganda (known as astroturfing) at three levels of government—central, provincial, and local.

LGBTQIA+ content sporadically hits levels of high political sensitivity. Fearing the threat of closure, internet companies comply with government censorship guidelines. Pre-emptive repression, both online and offline, frequently precedes anniversaries of occasions that are deemed politically sensitive, which Truex (2018) named the "dissident calendar." Repression strategies most commonly coincide with the five-year anniversaries of the Tiananmen Square Massacre, the founding of the PRC, and high-level CCP meetings.

> Every year before the 10-1 [National Day of the PRC], [the police] really put the people on the so-called list, such as people who had previously gone petitioning. Some people who have petitioned may not be able to leave their homes during that period, and there are people watching them. [At that time, the police] basically called me once a day, asking questions like: what kind of activities do you do? Are you mentally ill? Are you autistic? ... You can't do whatever you are doing either way. *(Wang Tie, genderqueer activist)*

> In 2019, [public security officers] approached me three or four times before and after the National Day. That's a lot, right? *(Xiaogang, gay activist)*

Indeed, the year 2021 that marked the centenary of the CCP (un)coincidentally saw a large-scale closing down of university LGBTQIA+ group social media accounts on WeChat (Bernot 2021). A person who has connections with staff working for WeChat commented:

> I tried to find some people [working for WeChat] to ask what exactly happened. It's very difficult to find any proof...after some investigation, our conclusion was that it might have been the company's action based on their own fear of getting into trouble. They wanted to prove to the authorities that they were loyal. *(Zhao Yun, queer woman activist)*

Automation increasingly supports social media content censorship—either by removing content or boosting anti-LGBTQIA+ content. The removal of content often follows the "dissident calendar" such that LGBTQIA+ activists have learned how to predict the nature of the content that will be made to disappear by observing common censorship patterns:

> [Social media censorship] is quite common in China, and everyone is used to it. Some sensitive topics are deleted on these kinds of public platforms. We're pretty used to it now. It's just an article. When a sensitive article is sent out, we must read it as soon as it is sent out and then take a screenshot to save it because it will be gone soon. *(Yun, trans activist)*

The boosting of anti-LGBTQIA+ content has been a more recent development in attempts to "organically" drown out queer voices from some online spaces. Some activists reported the double standard that is applied to the reporting function on popular social media platforms, most notably Weibo; while LGBTQIA+ content could be reported and then taken down, anti-LGBTQIA+ slurs are not taken down, even if reported. Indeed, recent scholarship has found that *who* is posting (not *what* is being posted) is often the trigger for censorship actions to co-opt, repress, and limit the reach of influential social media accounts not aligned with Party thought (Gallagher and Miller 2021).

> If you post some content around the stigmatization of LGBT [people] by sharing some popular science knowledge online, you will be blocked, and your reply will be deleted. But the slurs against the LGBT communities are here to stay. When other platform users report your account, your account will be closed, but when you report those who abuse you, they will be fine. *(Yueyue, trans activist)*

In the project, social media was often identified as a double-edged sword: interviewees viewed social media as an essential tool to help LGBTQIA+ communities connect, while simultaneously fearing that social media content could be erased at any time without notice, which could open their groups to surveillance, or attract hate online. Automation tools are used by platforms in various forms, such as reporting tools, select keyword/topic censorship, shadow banning, and astroturfing.

Surveillance and criminalization

An alarming finding from the dataset suggests that LGBTQIA+ groups and organizations are now under the constant scrutiny of both public security (公安) and state security (国安) in China. One wrong funder, social media post, or event can land an organization in hot water. The queer social sorting discussion above described the legal and regulatory context that allows such surveillance and repression to be institutionalized. Having set up the queer social sorting mechanism, the Party-state can assume the position of "virtuous Leviathan" (Lin and Trevaskes 2019, 41) in censoring queer voices; it can claim to be firmly grounded in established values (the return to "moralized" traditional gender roles) and acting within the bounds of established laws and regulations.

As discussed in the *queer social sorting* section, the Xi Jinping era of governance ushered in several laws and regulations that facilitated the precarity of LGBTQIA+ group activities (Wang 2021), potentially making those activities illegal. Those regulatory and legal changes now provide the imprimatur for public and state security agencies to monitor LGBTQIA+ groups; hence, stuck between a rock and a hard place, LGBTQIA+ groups have an ever-narrowing scope of activities that are deemed legal. As shown in the previous sections, formal organizational registration, fundraising, community organizing, and volunteer mobilization, among other activities, can now be framed as breaching legal and even criminal bounds set by the Party-state. These conditions of "illegalization and precarity" (Wang 2021) allow for the next step in surveillance and censorship to take place, that is, covert criminalization of LGBTQIA+ groups.

China's public security apparatus monitors reported crime as well as groups of people termed "target populations" (重点人口). This term has been used historically, including to refer to migrant worker populations in the 1980s. Under the guise of maintaining social stability, the surveillance and management of target populations was written into the job description of the Ministry of Public Security. Nowadays, the term "target populations" refers to those groups classified as risks to state security, those suspected of major criminal activity, general troublemakers, ex-convicts, and drug users; however, the criteria vary by location (Wang 2005). The attention of China's public security apparatus is consistently focused on target populations, while the growing state surveillance apparatus has made that surveillance task easier. An interviewee commented on feeling like he was on a close watchlist.

> The work mechanism of the police in China includes groups of people who are closely watched, so I am sure that, in the short term, I will definitely not be able to do anything else...I was mainly approached by the police in Guangzhou, and my colleagues were approached in other places. We don't have an office, ... we work online, and I'm the main person in charge. So, in our current situation, we can basically no longer operate. If there is any communication between us, the police will know and come looking for me, which would be more troublesome. *(Ding'An, gay activist)*

The findings of the interview suggest that, in some provinces, members of LGBTQIA+ groups may be placed on a perpetual watch list, similar to the way in which other target populations are monitored. Interviewees across several provinces reported instances where public security agencies sought them out to create lists of names of people involved in LGBTQIA+ activism. In one instance, an interviewee reported that they were invited to dinner (*read*: obliged to go) by public security officers and repeatedly questioned about the people who were likely to continue LGBTQIA+ organizing work following the closure of their organization *(Wang Tie, genderqueer activist)*. In another instance, a respondent's laptop, hard drive, and mobile phone were confiscated for several hours for the security agency to copy their content *(Ding'An, gay activist)*; the data was later used to develop a named network of all people involved with that organization *(Ding'An, gay activist)*. It is likely that, following the development of a group's network, all stakeholders will be added to a watchlist. There is probably a strong element of automation in the policing of such lists as they now feature numerous reported public security investigations into LGBTQIA+ groups, their members, and activities. Further research is needed to understand how such lists are used (such as whether they are used to make a permanent note on an individual's record).

Criminalization was also linked to the size of the city. The three megacities of Beijing, Guangzhou, and Shanghai, for example, consistently maintained public security surveillance of LGBTQIA+ groups, while evidence of surveillance and queer group criminalization was varied in other localities. The evidence from queer groups in smaller cities was mixed, some reporting that it was extreme, especially if their group was the only longstanding organization in the province; however, others noted that they were able to fly under the radar if they did not explicitly discuss politically-sensitive topics. A respondent who was a member of a China-wide organization commented on those differences:

> All big organizations in Beijing and Guangzhou are definitely investigated… In fact, most of the advocacy organizations are in Beijing and Guangzhou. Organizations in other places are primarily focused on AIDS prevention or social work, some are volunteer organizations without full-time staff, which is relatively mild [in terms of their political sensitivity]. *(Ding'An, gay activist)*

In addition to the above, the involvement of state security agencies in disrupting LGBTQIA+ group organizing confirms that the state security apparatus views LGBTQIA+ organizing as criminal in that it disrupts social stability (社会稳定). The Ministry of Public Security (公安) in China is responsible for public and political security and maintaining stability (维稳) in the country; the Ministry of State Security (国安) is the domestic intelligence and security agency that monitors and investigates national threats. The interview data confirms that LGBTQIA+ groups are monitored by both public and state security institutions. The elevation of LGBTQIA+ organizing from a public security concern to a state security issue was observed in the interview data in cases where fundraising and/or international collaborations were involved. Several interviewees reported on the intense process of going through an investigation:

> In 2018 we got into a little bit of trouble with the University administration… they talked to us and we drank tea. They requested twenty-three meetings with me over the course of three days and said the police had asked them to talk to me. I also received some calls from the state security department, but I didn't go to meet them. … They are still meeting my colleague regularly…they tell him not to go abroad first for his education, so almost he's banned from leaving the country. We didn't know because they didn't take his passport away…we could only find out that this happened when he tried to go [abroad]. *(Wangwang, trans activist)*

> My account was taken away by them, and last year, some of my friends had their computers and mobile phones confiscated. During that time, our work and personal information were copied, exposing many of us. It is especially unsafe. *(Xiaogang, gay activist)*

This section presents an exploratory glimpse into the potential criminalization of queer activists and advocates within the broader context of institutionalized queer social sorting. The data collected suggests that public and state security investigations rely on human intelligence officers who often aim to establish relationships with prominent activists to elicit information, such as by collecting lists of queer activists and advocates and keeping tabs on activities and funding. Surveillance via digital systems complements monitoring efforts by providing routine monitoring data. The involvement of state security investigations is primarily triggered by the involvement of foreign funding and foreign activists, indicating that the national focus is on eliminating "hostile foreign forces" (Zhao 2022).

Public security key performance indicators (KPIs)

The data analysis suggests that public security efforts at monitoring and controlling LGBTQIA+ people may increasingly treat them as a threat to social stability. In addition to clear work directives that now require public security officers to keep continuous watch over LGBTQIA+ groups, queer surveillance is also linked to automated public security systems. In the reports of the interviewees there are indications that public security officers can report surveillance and monitoring of LGBTQIA+ groups as their KPIs. As the examples in this section will show, such KPIs can seemingly be met by providing information that can be logged and placed on a public security watchlist. A previously unexplored perspective presented in this section demonstrates early evidence of these chilling practices.

Participants have reported being monitored by public security actors and understanding that sometimes all they needed to do was to "provide something" (such as organization activity details, plans, or activist names) to the officer in contact with them in order to be left alone. At times, the public and state security officer getting in touch with an LGBTQIA+ activist did not quite understand who LGBTQIA+ communities were, or what a particular group they were assigned to monitor was doing. This allowed the LGBTQIA+ organizers some space to withhold information or provide erroneous information in order to obstruct the investigation process. One participant, for example, reported being treated as an informant by a state security agent:

> He basically regards me as a source of information about sexual minority communities... Then this year this state security officer approached me, and asked, "Do you know the [the name of the organization]?"...He said it was "a new force" (新势力) in our gender circle. *(Xiaoyue, lesbian activist)*

> [The public security officers] invited me to a dinner when they decided to leave my city and they asked me to provide a list of names of who will continue the organization's activities. I

> just gave two or three random names to them, and then it was fine, they didn't come to me anymore, and there were no more requests for information. *(Wang Tie, genderqueer activist)*

In three interviews, the respondents referred to understanding that the public security stakeholders needed to meet their own KPIs, often without validating the information given.

> [Public security officers] do have KPIs, and the KPIs can be reported by their senior officers. The senior officers must have a plan, they must always have something planned. For example, finding key populations. LGBT [people] might be their target group. *(Wang Tie, genderqueer activist)*

One interviewee said that their parents were abruptly approached about their involvement in LGBTQIA+ activism when they were travelling abroad and were not approachable via a Chinese phone number, and that this was tied to possible KPI quotas:

> You know, maybe it was the end of the year or the middle of the year, and they have their KPIs to finish. *(Zhang Shang, trans activist)*

One respondent joked that she understood the importance of showing up to regular meetings with police officers and that she handled the stress of the meetings via a small act of rebellion—selecting Starbucks as the meeting place and ordering the most expensive thing on the menu for which the public security officer was obliged to pay *(Bo Yue, lesbian activist)*.

As the interviews show, there is an element of mediation between the KPIs of public security officers, the KPIs and policing quotas they are trying to reach, and the communities they monitor. At times, public security officers will reach out to LGBTQIA+ activists or their families, apparently to check a box on their to-do list. When surveillance of queer groups is escalated to the level of state security (usually when there is a breach of a law linked to funding and/or the involvement of foreign activists), the public security monitoring may continue in its current form or be taken over by state security officers.

This section argues that there is emergent evidence that public security actors have KPIs linked to the monitoring of LGBTQIA+ groups and their activities. Further research is needed to understand how the surveillance and monitoring systems of public security investigate LGBTQIA+ populations, and to comprehend the role that automation plays in organizing this public security work. This could be achieved by conducting in-depth interviews with LGBTQIA+ activists who have been investigated by public security actors, while carefully accounting for the sensitivity of such data.

Conclusion

Queer social sorting is emblematic of China's data-driven system of population management and social governance. Currently, LGBTQIA+ organizations face overlapping legal and regulatory restrictions that extend to key areas of activism and advocacy: fundraising, in-person event organizing, volunteer and staff recruitment, and organizational online presence. Additional pressures on organizations are added through the covert monitoring and control of the online and physical presence and activities of LGBTQIA+ organizations—a part of the everyday reality for LGBTQIA+ groups in most regions of China. Most LGBTQIA+ activists report being monitored and harassed by the police (公安), with some even reporting run-ins with state security authorities (国安). Like other activist groups, such as feminist or labor activists, queer groups have seen increased surveillance and repression efforts directed at them. Not fitting into the normative role agenda of the Party-state, LGBTQIA+ communities are being framed as a national security threat backed by "hostile foreign forces" (People's Daily 2022; Zhao 2022). Divided into three thematic parts, the chapter reviewed the broad political–ideological context, the framing of queer social sorting, and, finally, the strategies and tools used to implement surveillance and suppression of LGBTQIA+ groups on the ground.

The first section of the chapter explained the political–ideological context of China's call to return to more traditional gender roles that would help build the China Dream (中国梦). Such narrative signaling supports the regulatory and legal changes undertaken by the Xi Jinping administration: a suite of laws and regulations that have restricting effects on LGBTQIA+ groups and organizations. In the second section of the chapter, I explained how the dual repression of LGBTQIA+ organizations through restrictive laws and regulations, as well as covert monitoring, create an environment where routine activities of LGBTQIA+ groups in China can be monitored; they can easily become illegal and even criminal. An activist with over a decade of experience described this as feeling like a fish trapped in a fish tank—visible from the outside but completely restricted (Bernot and Davies 2023). Understanding the current "fish tank" environment in which LGBTQIA+ groups in China work is crucial in providing politically-sensitive support to those groups. Most importantly, there must be a shift away from encouraging advocacy activities that put people in further danger and towards more strategic support of local community activities through different means.

The final section of the chapter explored the strategies of public and state security agencies in LGBTQIA+ monitoring and investigation, paying particular attention to how automated decision-making is beginning to play a supporting role in state-supported surveillance. The findings of this research suggest that LGBTQIA+ groups are now under observation by increasingly automated systems of surveillance while subject to a blend of non-digital practices of public security surveillance (such as control through intimidation). The digital medium does not fundamentally change the conception of crime; rather, it reflects the social conception of crime via the digital medium (Stratton et al. 2017), shaping how monitored, illegal, and criminalized activities are surveilled. Automation now provides a toolkit for both public and state security agencies to monitor and control LGBTQIA+ groups, their networks, and their activities.

This chapter considered the alarming practices of public and state security surveillance investigation now directed at LGBTQIA+ groups. While China is maintaining political decorum, queer groups and organizations face continued control and active state-sponsored efforts at repression. This trend in LGBTQIA+ control and suppression has become clear during the Xi Jinping administration. Confronted by state-sponsored calls to return to traditional gender roles and clear strategies of queer social sorting, LGBTQIA+ groups are now being forced to close; the activities that such groups would typically engage in are being systematically outlawed. This chapter used a case study of LGBTQIA+ activism to show how China's experimentation with automation technologies of datafication and algorithmic governance has entrenched patriarchal and normative gender roles in the ordering of social life in China.

References

Bao, Hongwei. 2011. "'Queer Comrades': Transnational Popular Culture, Queer Sociality, and Socialist Legacy." *English Language Notes* 49 (1): 131–137. https://doi.org/10.1215/00138282-49.1.131.

Bernot, Ausma. 2021. "China's Forced Invisibility of LGBTQ Communities on Social Media." *The Interpreter*, July 9, 2021. https://www.lowyinstitute.org/the-interpreter/china-s-forced-invisibility-lgbtq-communities-social-media.

Bernot, Ausma. 2022. "Double-Speak as LGBTQI+ Resistance." in *China Story Yearbook 2022: Chains* edited by Linda Jaivin, Esther Sunkyung Klein and Annie Luman Ren, 199–204. Canberra: ANU Press.

Bernot, Ausma and Sara E. Davies. 2023. "The 'Fish Tank': Social Sorting of LGBTQIA+ Activists in China." *International Feminist Journal of Politics*, Online First. https://doi.org/10.1080/14616742.2023.2261948

Bernotaite, Ausma, H. C. Zhuo and Lukas Berredo. 2017. "Voices from Trans Communities in China: Summary Report of Three Consultations." *Asia Catalyst*, 2017. https://archive.org/details/2017-voices-from-trans-communities-in-china/mode/1up.

Chappell, Bill. 2018. "Weibo Bans Gay Content — And Quickly Reverses Itself After An Outcry." *NPR*, April 16, 2018. https://www.npr.org/sections/thetwo-way/2018/04/16/602902197/weibo-bans-gay-content-and-quickly-reverses-itself-after-an-outcry.

Chia, Joy L. 2019. "'What's Love Got to Do with It?': LGBTQ Rights and Patriotism in Xi's China Special Issue: Legal Regimes of Sexual Orientation and Gender Identity in Asia." *Australian Journal of Asian Law* 20 (1): 27–38. https://heinonline.org/HOL/P?h=hein.journals/ajal20&i=28.

Creemers, Rogier. 2015. "The Pivot in Chinese Cybergovernance: Integrating Internet Control in Xi Jinping's China." *China Perspectives* 2015 (4): 5–13. https://doi.org/10.4000/chinaperspectives.6835.

Gallagher, Mary and Blake Miller. 2021. "Who Not What: The Logic of China's Information Control Strategy." *The China Quarterly*, 248 (1): 1011–1036. https://doi.org/10.1017/S0305741021000345.

Gu, Yijia and Luke Heemsbergen. 2023. "The Ambivalent Governance of Platformed Chinese Feminism under Censorship: Weibo, Xianzi, and her Friends." *International Journal of Communication*, 17: 3822–3843.

Hildebrandt, Timothy. 2012. "Development and Division: The Effect of Transnational Linkages and Local Politics on LGBT Activism in China." *Journal of Contemporary China*, 21 (77): 845–862. https://doi.org/10.1080/10670564.2012.684967.

Jeffreys, Elaine. 2017. "Public Policy and LGBT People and Activism in Mainland China." In *Routledge Handbook of the Chinese Communist Party* edited by Willy Wo-Lap Lam, London: Routledge.

Kang, Wenqing. 2009. *Obsession: Male Same-Sex Relations in China, 1900–1950*. Hong Kong: Hong Kong University Press.

Kellogg, Thomas E. 2012. "Western Funding for Rule of Law Initiatives in China: The Importance of a Civil Society Based Approach." *China Perspectives*, 3: 53–59. https://journals.openedition.org/chinaperspectives/pdf/5954.

Kenyon, Miles. 2020. "WeChat Surveillance Explained." *The Citizen Lab*, May 7, 2020. https://citizenlab.ca/2020/05/wechat-surveillance-explained/.

King, Gary, Jennifer Pan and Margaret E. Roberts. 2013. "How Censorship in China Allows Government Criticism but Silences Collective Expression." *American Political Science Review* 107 (2): 326–343. https://doi.org/10.1017/S0003055413000014.

Kunlun Ce Research Institute. 2023. "'LGBT' is about China's Non-Traditional National Security." May 19, 2023. https://archive.md/wip/H05M8

Lin, Delia and Susan Trevaskes. 2019. "Creating a Virtuous Leviathan: The Party, Law, and Socialist Core Values." *Asian Journal of Law and Society*, 6 (1): 41–66. https://doi.org/10.1017/als.2018.41.

Lindberg, Frida. 2021. "Women's Rights in China and Feminism on Chinese Social Media." *Institute for Security and Development Policy*, June 2021. https://isdp.eu/publication/womens-rights-in-china-and-feminism-on-chinese-social-media/.

Lyon, David. 2002. "Surveillance as Social Sorting: Computer Codes and Mobile Bodies." In *Surveillance as Social Sorting* edited by David Lyon, 13–30. London and New York: Routledge.

Miller, Blake. 2018. "Delegated Dictatorship: Examining the State and Market Forces behind Information Control in China." PhD diss., University of Michigan.

Nardi, Dominic J. 2019. "Country update: China. Religious Freedom in China's High-Tech Surveillance State." *United States Commission on International Religious Freedom*, September 2019. https://www.uscirf.gov/countries/china/religious-freedom-chinas-high-tech-surveillance-state.

People's Daily. 2022. "Let's Look at the Five Deadly Crimes of the 'Five Eyes Alliance'." September 27, 2022. https://archive.md/wip/aet13

Plantin, Jean-Christophe and Gabriele de Seta. 2019. "WeChat as Infrastructure: The Techno-Nationalist Shaping of Chinese Digital Platforms." *Chinese Journal of Communication* 12 (3): 257–273. https://doi.org/10.1080/17544750.2019.1572633.

Qiang, Xiao. 2019. "The Road to Digital Unfreedom: President Xi's Surveillance State." *Journal of Democracy* 30 (1): 53–67. https://heinonline.org/HOL/P?h=hein.journals/jnlodmcy30&i=51.

Qiao, Leshui. 2022. "Mainland China's TERFs' Misogyny Under JK Rowling's Anti-trans Incident." *Proceedings of the 2022 8th International Conference on Humanities and Social Science Research (ICHSSR 2022)*. https://doi.org/10.2991/assehr.k.220504.238.

Ren, Annie Luman. 2022. "How the 'Garlic Chives' Grieved: A Song for China's Three-Child Policy." In *China Story Yearbook 2021: Contradiction*, edited by Linda Jaivin, Esther Sunkyung Klein and Sharon Strange, 171–177. Canberra: ANU Press. https://doi.org/10.22459/CSY.2022.05A.

Ren, Xiaoyi and Tianhan Gui. 2022. "Where the Rainbow Rises: The Strategic Adaptations of China's LGBT NGOs to Restricted Civic Space." *Journal of Contemporary China*: 1–19. https://doi.org/10.1080/10670564.2022.2131378.

Ruskola, Teemu. 1993. "Law, Sexual Morality, and Gender Equality in Qing and Communist China Note." *Yale Law Journal*, 103 (8): 2531–2566. https://heinonline.org/HOL/P?h=hein.journals/ylr103&i=2545.

Shen, Caoyang. 2023. "Where is a Safe Online Home? Challenges Faced by Chinese Queer Communities in Speaking Out on Douyin." Master's diss., University of Michigan.

Stratton, Greg, Anastasia Powell and Robin Cameron. 2017. "Crime and Justice in Digital Society: Towards a "Digital Criminology"?" *International Journal for Crime, Justice and Social Democracy*, 6 (2): 17–33. https://doi.org/10.5204/ijcjsd.v6i2.355.

The State Radio and Television Administration. 2021. "Notice of the General Office of the State Administration of Radio and Television on Further Strengthening the Management of Cultural Programs and Their Personnel." *China Law Translate*, September 2, 2021. https://www.chinalawtranslate.com/en/content-regulation/

Truex, Rory. 2018. "Focal Points, Dissident Calendars, and Preemptive Repression." *Journal of Conflict Resolution*, 63 (4): 1032–1052. https://doi.org/10.1177/0022002718770520.

Wang, Fei-Ling. 2005. *Organizing through Division and Exclusion: China's Hukou System*. Stanford University Press.

Wang, Stephanie Yingyi. 2021. "Unfinished Revolution: An Overview of Three Decades of LGBT Activism in China." *Made in China Journal*, 6 (1): 90–95. https://doi.org/10.22459/MIC.06.01.2021.11.

Wang, Stephanie Yingyi. 2023. "Fare Thee Well Beijing LGBT Centre." *Made in China*, June 8, 2023. https://madeinchinajournal.com/2023/06/08/fare-thee-well-beijing-lgbt-centre/#:~:text=One%20of%20the%20largest%20and,young%20generation%20of%20queer%20activists.

Wilhelm, Katherine. 2022. "Chinese LGBT rights activist speaks at USALI." *U.S.-Asia Law Institute*, April 26, 2022. https://usali.org/institute-news/chinese-lgbt-rights-activist-speaks-at-usali.

Worth, Heather, Jing Jun, Karen McMillan, Su Chunyan, Fu Xiaoxing, Zhang Yuping, Rui Zhao, Angela Kelly-Hanku, Cui Jia and Zhang Youchun. 2018. "'There Was No Mercy At All': Hooliganism, Homosexuality and the Opening-up of China." *International Sociology*, 34 (1): 38–57. https://doi.org/10.1177/0268580918812265.

Xinhua News. 2021. "Xi Jinping on Family Values." May 5, 2021. https://archive.md/rLJCD

Yang, Rujun. 2023. "Mosaic of Beliefs: Comparing Gender Ideology in China across Generation, Geography, and Gender." *International Journal of Comparative Sociology*. Online First. https://doi.org/10.1177/00207152221147493.

Yang, Yue. 2021. "When Positive Energy Meets Satirical Feminist Backfire: Hashtag Activism during the COVID-19 Outbreak in China." *Global Media and China*, 7 (1): 99–119. https://doi.org/10.1177/205943642110213.

Yin, Siyuan and Yu Sun. 2021. "Intersectional Digital Feminism: Assessing the Participation Politics and Impact of the MeToo Movement in China." *Feminist Media Studies*, 21 (7): 1176–1192. https://doi.org/10.1080/14680777.2020.1837908.

Yu, Haiqing. 2016. "The Biopolitics of China's HIV Governance." In *New Mentalities of Government in China*. David Bray and Elaine Jeffreys. London, Routledge. https://doi.org/10.4324/9781315688848.

Zhao, Yuxin. 2022. "Waves Under the Surface: The LGBTQ NGO Field and its Cultural Changes in China." Master's diss., University of Chicago.

Zhou, Xiaoyao, Jingjing Ye, Hao Li and Haiqing Yu. 2022. "The Rising Child Penalty in China." *China Economic Review*, 76 (101869): 1–16. https://doi.org/10.1016/j.chieco.2022.101869.

CHAPTER 9

Regulating Algorithmic Price Discrimination on Chinese Digital Platforms

Haiqing Yu and Xuanzi Xu

Abstract

This chapter examines the process in China's effort to regulate algorithmic price discrimination on digital consumer platforms. It pays attention to the role of social actors—consumers, consumer associations, academics and legal scholars, and state media—in the participatory process of Chinese public policymaking on algorithms regulation. We argue that regulating algorithmic price discrimination exemplifies China's dual-tracked and tiered approach to algorithm and platform governance. Algorithms governance is not just about regulating algorithmic platforms but also social governance through algorithms via platforms. Such a governance framework will reshape how the technology is built and deployed in and beyond China, impacting Chinese technology exports, the outbound platform economy, and global AI governance.

Keywords: Algorithmic price discrimination; Algorithmic regulation; Platform governance; China

Introduction

Algorithmic price discrimination, known in Chinese as *dashuju* (big data-enabled 大数据) *shashu* (price discrimination against existing customers 杀熟),[57] has been discussed on Chinese social media platforms, in the media, and in scholarly publications since 2017. It is a kind of algorithmic pricing personalization that supposedly goes through an automated decision-making process and assigns higher prices to users who are usually loyal customers willing to pay, and lower prices (with big discounts) to new and infrequent users for the same product or service, based on user profiles and historical data on the platforms. Such data encompasses everything from consumption records (frequent users or new users, for example) and geographic data (such as rich or poor place of residence) to communication habits (whether iPhone or Huawei phone users, for instance), and even to biodata

[57] The Chinese term *shashu* has predated the digital era, as pointed out later. For the rest of the chapter, the Chinese term is used with quotation marks only when referring to its cultural and historical significance.

captured by surveillance cameras in shops and other public places. Quite often, loyal customers and iPhone users would pay more than new customers and Android users for services such as food delivery, hailing of taxis, and bookings on travel platforms.

Algorithmically personalized pricing or dynamic pricing is a worldwide practice across various sectors, from the airline and insurance industries to online consumer retail markets. Internet and tech companies are well-known for using data analytics (based on large volumes of consumer data collected from their proprietary platforms) to recommend personalized content, enact automated content moderation, and practice personalized pricing in order to increase market outreach and maximize profit. As noted by Moor and Lury (2018, 502), "contemporary pricing practices bring together individualizing and dividualizing practices in ways that have significant implications for processes of discrimination, identification, and collective action." Algorithmic price discrimination is a step further toward algorithmic bias than dynamic pricing, as the latter adjusts prices according to supply and demand in real time and place (as with Uber rides), while the former adjusts prices based on five personalized dimensions: *historical* (history and frequency of using the service or platform), *temporal* (time, day, and month), *spatial* (location and locality), *relational* (social status and networks), and *material* (device brands and operation systems). Algorithmic price discrimination exploits and abuses such personal data.

Chinese platform companies are not averse to using the algorithmic pricing practices that "maximise profit and influence". In fact, they have optimized the "*shashu*" tradition in Chinese societies in which business activities and transactions often start in one's family and social networks. Trading among known people in one's social networks and among frequent customers can reduce social anxiety and the cost of trust-building for start-ups and small businesses. Price discrimination against existing customers started to appear in the 1990s when some traders began to exploit both people in their own social networks and loyal customers, as in the notorious pyramid schemes. It has continued to grow despite government crackdowns (Chow 2018). In such a model, traders and buyers know and have direct communication with one another through commonly-shared social networks.

Big-data-driven algorithmic price discrimination via digital platforms does not have the personalized structure such as that between traders and buyers. It operates on hierarchical and opaque systems, with traders hiding behind platform algorithms and buyers being "transparent" to the former through platform data analytics provided to traders. Its pricing techniques are automated by algorithmic optimization software and distributed ledger technologies that are hidden and beyond the control of consumers and users. Because of the unequal trading relationships, civil society groups, legal scholars, media, and antitrust regulators have called for better protection of consumer rights and management of market competition.

Amid the call for greater platform regulation and algorithmic transparency, Chinese regulators have acted quickly to rein in the power of algorithmic platforms. Algorithms, defined as "automated decision-making" in the "Personal Information Protection Law" (PIPL, enacted on 20 August 2021), are seen by Chinese lawmakers as central to the operation of Internet platforms and responsible for the "information cocoons" created by these platforms through recommendation systems; as such, they must be regulated to ensure transparency, liability, accountability, and fairness by platforms and to protect consumer rights.

China is among the first countries to regulate artificial intelligence (AI) algorithms. The new algorithmic regulation took effect in March 2022, requiring businesses to provide explainable AI algorithms and be transparent when recommending products or services. The new regulations prohibit businesses reliant on AI algorithms from offering different prices to different people; they are also required to notify users, and allow them to opt out, when algorithms are used to make recommendations. Companies that violate the rules could face fines, be barred from registering new users, have their business licenses revoked, or see their websites or apps shut down. Such algorithmic regulation came after more than three years of heated public discussion, numerous complaints to consumers' associations, and a number of publicized legal cases.

Over four years (2018–2021), we followed social media posts referencing algorithmic price discrimination, mainly on Weibo (microblogging) and Zhihu (Q&A forum); we examined published news stories, government announcements, legal documents, and survey reports—all available in the public domain.[58] Based on such a longitudinal analysis of publicly available data and using critical discourse analysis,[59] this chapter examines the process in China's effort to regulate algorithmic price discrimination on digital consumer platforms. It pays attention to the role of social actors—consumers, consumer associations (as government organized non-governmental organizations, or GONGOs), academics and legal scholars, and state media—in the public policymaking process and the push-and-pull battle

[58] Data collection stopped at 2021 because of COVID-19: mobility and travel were restricted from 2021 to 2023 and people's attention turned to survival imperatives instead of prices or price discrimination.

[59] The data analysis in this chapter is based on a selected pool of 957 social media posts, 60 news stories from official media, 14 government announcements and legal documents, and four survey reports from the China Youth Daily (2018), Chinese Consumers Association (2019), Fudan University (2020), and Southern Metropolitan News (2020). We took an iterative approach to content analysis, paying attention to major themes and discourses, rather than who said what. For lack of space, we cannot elaborate on methods.

between big tech and big government. It identifies key players and events that stirred up public discussion about discriminatory pricing practices. The aim is to not only map out the contours of public opinion on the issue, but also to examine the interaction among different players and stakeholders in the agenda setting which eventually led to the policy and legislative changes.

There has been much written on algorithmic regulation and governance from legal and public policy perspectives. Researchers have explored legal issues in relation to data protection, consumer protection, anti-discrimination law, and anti-trust law (for example, Danaher et al. 2017; Ulbricht and Yeung 2022; Katzenbach and Ulbricht 2019). But how do media scholars make sense of the social and political implications of algorithms and their governance? What does the process of regulating algorithmic price discrimination tell us about China's public policymaking process? After all, price is "a fundamental yet obscure topic within media and communication studies" while it also "has both critical and practical value for current debates about digital distribution" (Lobato 2021, 315).

We view the process of regulating algorithmic price discrimination as a feature of China's emerging regime in algorithm and AI governance. As this chapter will discuss later, algorithmic governance is not just about the governance of algorithms; it also concerns social governance through algorithms via platforms. Such a governance framework will reshape how the technology is built and deployed in and beyond China, affecting Chinese technology exports, the outbound platform economy, and global AI governance.

The chapter starts with an analysis of Chinese social media posts by victims of algorithmic price discrimination, focusing on public discussion of prominent cases to illustrate how algorithmic price discrimination became a public topic and gained traction, from social media through to mainstream media. It then discusses how key social actors, from news media and consumer associations to industry governing bodies and local governments, joined the public discussion of algorithmic price discrimination, leading to new laws and regulations to govern the digital consumer market. Finally, it considers the social and political implications of algorithmic regulations on China's platform governance, furthering the debate on the significance of such platform governance for Chinese authoritarianism.

From Murmuring to Making Headlines

Complaints about algorithmic price discrimination started to surface on Chinese social media platforms from late 2017. Most were directed toward e-commerce, travel and hotel bookings, and taxi-hailing platforms. It started as trickles of individual complaints and grew into cascades of online public opinion as more cases were reported in the mainstream media. It gained further momentum from 2021 when a few legal proceedings were initiated by victims.

In December 2017, a Weibo user called "Master Liao" ("Liaoshifu liaoshifu") shared a post about his experience of being unfairly treated by a hotel booking platform and a taxi ride-hailing platform.[60] He complained that being a frequent user of the first platform and a paid VIP member of the second platform he was made to pay more for the same services than new or non-loyal customers. He lamented that consumers expected better services when giving up personal data and privacy to platforms, but instead found themselves victims of their own generosity at the "evil hands" of internet companies. This post immediately gained public attention, very soon attracting thousands of comments, and was reposted by over 20,000 users. Two months later, in February 2018, his story was reported in the *Science & Technology Daily*, the official newspaper of the Ministry of Science and Technology (Zhai 2018). This then drew further attention from other media outlets, such as the *New Evening Post* and *Guancha.*[61] These news stories in mainstream media were then added to official news portals, such as www.sina.com.cn and www.ifeng.com. Weibo accounts of influential mainstream media also joined the public discussion of price discrimination.

The acceleration of individual consumers' online murmuring into a trending topic on social media was achieved via the intervention of the mainstream media. The noise soon died down, but the momentum was picked up by other victims on various social commerce platforms. Stories were shared on Zhihu, Hupu (Gugugu 2018), Douban, and Weibo (Yu 2019), with more social commerce platforms exposed by users for their price discrimination practices. In 2018, "*shashu*" was included among the top 10 consumer violation cases ranked by state agencies, including the China Consumers Association (CCA).[62]

Weibo became the main platform on which people shared personal experiences of algorithmic price discrimination. They lodged complaints against platforms and private businesses via Black Cat Complaint, a website and app for consumers to file complaints, seek legal information, and find mediation services. Black Cat was established by Weibo's parent company, Sina, in March 2018, as a public hub for consumers to lodge complaints and for companies to address complaints. On the Black Cat platform, "*shashu*" quickly became a high frequency term, with complaints mainly about travel (ticket and hotel bookings), taxi-hailing, and online shopping platforms. Surveys by the Beijing Consumer Association in 2019 and the Southern Metropolitan News in 2020 showed that the majority of those surveyed had

[60] https://weibo.com/1644114654/G5tswEmqx?refer_flag=1001030103_

[61] https://www.sohu.com/a/224646862_338398; https://user.guancha.cn/main/content?id=6718

[62] http://www.hxnews.com/news/gn/gnxw/201901/25/1697836.shtml

experienced algorithmic price discrimination in everyday life and were strongly against such practices (Wang and Zhao 2019; Zhang, 2020). In September 2020, a report issued by Black Cat further illustrated the widespread practice of algorithmic price discrimination alongside increasing consumer awareness and intolerance of such practices (The Paper 2020).

Apart from making complaints on social media and consumer dispute resolution platforms, individual citizens have also sought to protect their rights and interests through legal proceedings. In July 2021, a Ms Hu from Shaoxing (Zhejiang province) sued the tourist platform Ctrip for price discrimination. Hu was a diamond member of Ctrip with a record of expenditure of over 100,000 RMB. In 2020 she paid RMB 2,889 per night for a luxury room in a five-star hotel, but found the same room was advertised at the hotel for RMB 1,377.63. She lodged a complaint to Ctrip, which only agreed to take partial responsibility; it merely refunded her the price difference. Ctrip blamed third party retailers on its platform for the price discrimination incident. Ms Hu wanted full platform liability in such practices, disputing Ctrip's consumer service and privacy agreements. She won the case on the grounds of Ctrip's dishonesty (in advertising), fraud (in collecting and misusing excessive and inessential consumer data), and failure to protect consumer rights, even when it was the third-party retailers on the platform who exploited consumer data for profiteering. Ctrip was made to triple the compensation it paid to Ms Hu and ordered to give Ms Hu the option to disagree with its consumer privacy agreement and opt out of its algorithmic recommendation systems (Classic case 2021). Ctrip has since modified its algorithms and enabled users to check and compare prices for different levels of membership (Li 2021b).

The 2021 Hu vs Ctrip case is not the first legal case in which consumers have sued platforms for price discrimination practices. In an earlier case (in 2019), a Mr. Liu sued Meituan (China's leading e-commerce platform for group-buying services, including food delivery) for charging him one yuan more than his colleague for the same delivery service. He lost the case because the judge ruled that platforms had the right to adjust delivery fees based on market demand at different times of the day, and that Liu did not provide sufficient evidence to support his claim (Wu 2020). His circumstances were seen as reflecting a regular practice in dynamic pricing, rather than a case of algorithmic price discrimination, and his case is less known than the Hu vs Ctrip case which has been widely discussed on Chinese social media platforms as a successful case in consumer rights protection.

The two examples (Master Liao on Weibo and Hu vs Ctrip) illustrate the key issue in public discussion of algorithmic price discrimination: consumers are treated as marketable entities and data subjects. There is little related discussion about platforms' monopoly over consumer data (from data harvesting to data appropriation). Reflecting the popular discourse on the topic, scholars and legal experts do not argue against the use of consumer data and algorithms by platforms and internet companies to deliver services through recommendation systems or to maintain their competitive advantages in market operations; rather, they argue against a phenomenon still confined to consumer affairs, that is, private companies using big data and algorithms to maintain their market monopoly provided they do not cause harm to middle-class consumers.

Scholars and legal experts engage in the debate via their own social media accounts and mainstream news media outlets. They support and amplify netizen murmurings and disgruntlement about algorithmic price discrimination; they advocate sweeping rules to regulate algorithms and AI, lobby law and policymakers, and contribute directly to the lawmaking process. A Fudan University research group led by Professor Jinyun Sun, for example, conducted a longitudinal survey from 2017 to 2020 on mobile taxi ride-hailing apps. It provided scholarly evidence that loyal customers paid more than new customers and that iPhone users were likely to be charged more than Android users (PDSM 2021). This research attracted wide coverage from the mainstream media, including the *Guangzhou Daily* (Feng, Zhang and Li 2021) and the *People's Daily* (People Net 2020).

Apart from providing scholarly reports, legal scholars have proposed changes to Chinese laws governing litigation, digital rights, and algorithms. They argue against the burden of proof being placed on consumers in litigations and for platforms and private companies being required to provide proof of innocence; instead, they propose public interest litigation to deal with price discrimination-related cases to better protect consumers' interests (Dong et al. 2021; Zhao 2019). They have suggested borrowing concepts from elsewhere, such as "the right to be forgotten" in the European Union's GDPR, to give Chinese consumers more control of their personal data, particularly to have the ability to control its inappropriate harvesting, misuse, and abuse by Chinese platforms (Zhang 2021). They also urged local authorities to develop new laws to regulate data and algorithms in a coordinated and flexible way, considering regional differences in the digital economy and therefore differences in requirements for data protection when developing the data industry.

Some leading scholars have contributed directly to drafting new laws and offered feedback to draft laws, such as the E-commerce Law (Zhao 2018) and the PIPL (Privacy Information Protection Law) (Jin 2021; Zhang T 2021). In the public consultation process for the draft PIPL, for example, legal experts proposed that big platforms should work with independent supervision teams to monitor users' personal information protection and regularly publish social responsibility reports (Jin 2021). They also suggested introducing the idea of "rights to data portability" to the law (Zhang T 2021). All those suggestions were adopted in the final version of PIPL. The revised Anti-Monopoly Law (AML, effective from August 1, 2022) is another case in point. In the new AML, provisions against the use of data and algorithms by platform operators with a dominant market position were added at the suggestion of legal scholars from multiple universities, led by legal scholar Professor Song Yang, deputy director of the Legal Committee of the People's Congress of Liaoning Province, during the consultation process for the draft law in 2021 (Li 2021a).

It is worth noting that leading public intellectuals in China have a symbiotic relationship with the government and are part of the tradition of scholar-officials in imperial China (Finger 2016). They occupy elite positions at the top of the social hierarchy, enjoying prestige, wealth, and power. They are favored by the media, mostly respected by the public, and as such carry a considerable weight of authority when engaging in policy consultation and lawmaking processes. Together with the netizens, they have turned murmurings and trickles of complaints into cascades of public opinion. The state media have also played a key role in amplifying the voice of the people, leading to behavioral and policy changes in algorithmic governance.

From Public Opinion to Public Policy

News media, consumer associations, industry governing bodies, and local governments have all played important roles in the move to regulate platforms and algorithms. News media are active agenda-setters on algorithmic price discrimination by platforms. Their reporting of online public discussions and offline consumer litigation has helped keep the momentum of public discussion on the topic in the public sphere. They interview legal scholars and report their findings, and produce investigative reports on algorithmic price discrimination based on consumer surveys. *China Youth Daily* was the first to publish its own survey report on March 15, 2018 (China's consumer rights day) when "*shashu*" emerged as a trending topic on social media. The survey showed that 51.3% of participants had experienced price discrimination, 59.2% felt helpless when facing powerful platforms, and 59.1% asked for better regulation (Du, Zhang and Qu 2018). It has since set an example for other news media to pursue the topic during the week of consumer rights day, creating the momentum for public policy intervention.

Local and national consumers' associations are key players in consumer rights protection. They have been active in engaging with news media, digital platforms, legal experts, and government officials to address thousands of complaints lodged by consumer victims of *shashu*. In early 2019, for example, the Beijing Consumers' Association publicized a survey report in mainstream media, including in the *People's Daily* (Zhao Y 2019), *Beijing Evening News* (Wang 2019), *Beijing Daily* (Zhao 2019), *Beijing Youth* (Xinhua Net reprinted) (Wang and Zhao 2019), and *Beijing Business Today* (Shao 2019). The news reports called for reinforcing regulatory and legal protection of consumers' personal data, improving governmental supervision mechanisms, enhancing private corporations' self-discipline, and raising consumer awareness regarding protection of their rights.

The national agency CCA also stepped in to convene meetings with key stakeholders to pressure private businesses to change their behavior and to urge government agencies to better regulate the industry. In one instance, on January 7, 2021, CCA convened a national meeting to discuss how to better regulate e-commerce-related algorithms and protect consumers' rights (CCA 2021). The meeting was attended by legal experts, local court representatives, information technology (IT) experts, lawyers, representatives of local consumers' associations, consumer representatives, state media, and key government agencies—including the Supreme People's Court, State Administration for Market Regulation (SAMR), Cyberspace Administration of China (CAC), and State Taxation Administration (STA). At the multilateral meeting, the CCA openly criticized digital platforms for breaching consumer privacy and misusing consumer data and suggested establishing a formal procedure to review complaints about algorithmic price discrimination under the supervision of independent technical appraisal agencies. One month later, in February, the "Anti-Monopoly Guidelines for Platform Economies" was issued by SAMR. Then, in April 2021, SAMR, together with CAC and STA, convened a meeting with 34 platforms, all dominant players in their sectors, to warn against monopolistic practices, including algorithmic price discrimination (SAMR 2021). Local branches of CCA all responded to Beijing's call to regulate the consumer market by engaging with their local government agencies, media, and legal experts, initiating policy responses from the bottom up.

Since 2019, a swath of laws, regulations, and provisions have been issued by national, provincial, and local governments and regulatory authorities to tighten data regulations in China (Table 9.1). The multiple provisions and regulations take from months to years (as with the AML) to develop and pass, with involvement of various interest groups. They all aim to regulate how (and how much) personal data is collected, used, stored, and traded to ensure data transparency, accountability, and fairness. Big tech companies are tasked with a gatekeeper responsibility in consumer data protection, beyond algorithmic price discrimination (Lu 2021).

Industry regulators and local governments acted first in response to the call for better and tougher regulations against misuse of consumer data for corporate profit. On 20 August 2020, the Ministry of Culture and Tourism (MCT) promulgated the "Interim Provisions on the Administration of Online Tourism Business Services," the first among Chinese ministries to take measures to regulate the tourism industry, the most notorious for discriminatory pricing practices and with most consumer complaints. In March 2021, the "Zhejiang Fair Online Trading" platform went online as a portal for supervising consumer-related digital transactions on major online trading platforms. Developed by the Zhejiang branch of SAMR, it was aimed at regulating anti-monopoly behaviors. On June 29, 2021, Shenzhen passed its "Data Law of Shenzhen Special Economic Zone" which imposes a fine of up to 5% of the total sales revenue of offending operators for engaging in algorithmic price discrimination practices. It also covers the provision of public interest litigation for victims, while its penalty rate (5% of total sales revenue) is much higher than the 0.1%–0.5% rate imposed on platform operators for discriminatory pricing practices stipulated by SAMR in its revision draft of the "Provisions on the Administrative Punishment of Price-related Violation," released on July 2, 2021.

Table 9.1 Selected laws and regulations that impact on *shashu* (compiled by authors)

Time	Law and regulation	Issuing body
1 Jan 2019	E-commerce Law	National People's Congress (NPC) Standing Committee
1 Oct 2020	Interim Provisions on the Administration of Online Tourism Business Services	MCT
7 Feb 2021	Anti-Monopoly Guidelines for Platform Economies	Anti-monopoly Commission of the State Council
Mar 2021	Zhejiang Fair Online Trading	Zhejiang Provincial Government
29 Jun 2021	Data Regulation of Shenzhen Special Economic Zone	Shenzhen Municipal Government
2 Jul 2021	Provisions on the Administrative Punishment of Price-related Violation	SAMR
1 Sept 2021	Data Security Law	NPC Standing Committee
1 Nov 2021	Personal Information Protection Law	NPC Standing Committee
1 Mar 2022	Internet Information Service Algorithmic Recommendation Management Provisions	CAC, SAMR, Ministry of Industry and Information Technology, Ministry of Public Security, etc
1 Aug 2022	Anti-Monopoly Law	NPC Standing Committee

The mainstream news media, together with consumers' associations and legal scholars, have played advocacy roles as instigators of new regulations, gatekeepers of draft legislation, and watchdogs for new legislation in implementation and compliance. On the issue of algorithmic price discrimination, the state media have joined legal experts to openly criticize new regulations and provisions for their lack of teeth. The *Guangzhou Daily* criticized the 2020 "Interim Provisions on the Administration of Online Tourism Business Services" for its weak compliance mechanisms and its lack of teeth for punishing algorithmic price discrimination practices that are explicitly forbidden (Feng, Zhang and Li 2021). Xinhua News Agency pointed out the flaws in multiple regulations regarding protecting consumers from algorithmic price discrimination (Dong et al. 2021).

These public actors played a particularly important role in the public consultation process. The E-Commerce Law went through three reviews in 2017 and 2018. A provision on algorithmic price discrimination was added to the third draft as the topic started to trend on Chinese social media in 2018 (Bai et al. 2018). Article 18 of the 2019 E-commerce Law stipulates that internet users should have the opt-out option on e-commerce platforms in product or service recommendations. Although it is touted as protecting consumer rights while highlighting the responsibilities of service/product providers and the platforms, the law was criticized for its weakness by the *People's Daily*—citing legal experts—for its setting of subjective and ambiguous parameters around algorithmic manipulation by platforms, and therefore lacking the power to protect consumer rights (People Net 2020).

Similarly, article 24 of the PIPL (specifically on "*shashu*") went through two revisions before going into effect. The revisions were made via the feedback process, mainly by lawmakers, local governments, governmental departments, and legal scholars (Zhang 2021). Such a consultation process was also seen in the revision of the Anti-Monopoly Law. The 2008 Anti-Monopoly Law was criticized for its failure to regulate price discrimination, despite its prohibition in writing, as the law was only applicable to "oligopolistic platforms" without clear criteria on how to define oligopoly and oligopolistic practices, while algorithmic price discrimination had been a pervasive problem on all platforms. In response, the amended 2022 Anti-Monopoly Law prohibited the use of "data and algorithms, technology, or capital advantages and platform rules" to engage in anti-competitive behavior (Article 9) and increased fines, liability, and damage to credit records for violations (Articles 56–67) (Huld 2022).

China has also issued a suite of regulations on algorithms and AI, including regulations on recommendation algorithms, effective from March 1, 2022 (CAC et al. 2021—for an English translation, see Creemers, Webster and Toner 2022). The regulations focus on algorithm-related security matters and require internet platforms to register algorithm services in a special action called the "Qinglang Algorithm Integrated Management." Internet companies and developers are required to file with the algorithm registry, a mandatory registration and repository system newly established by China's Internet regulators for recommendation algorithms (https://beian.cac.gov.cn/#/index). Some of China's biggest internet platform companies, including Tencent, Alibaba, Bytedance, Meituan, and JD have all filed their algorithms with the registry.

The algorithmic recommendation regulations have followed the principle of "differentiated regulation and management" via classification and gradings.[63] They cover provisions that outlaw algorithmic price discrimination and restrict tech companies' usage of algorithmic recommendations on major ride-hailing, e-commerce, streaming, and social media companies. The regulations stipulate that tech companies must inform users "in a conspicuous way" if algorithms are being used to push content to them. Users will be allowed to opt out of being targeted with algorithmic recommendations.

Chinese state media have mostly cooperated with the regulatory authorities and gone into full propaganda mode to celebrate the regulatory measures that protect consumer rights. Two weeks after the algorithmic regulation took effect, Chinese Central Television issued its annual special report on consumer rights in China on the national consumer protection day, March 15, 2022, staging a gala powered by AI and AR (augmented reality) technologies to promote consumer awareness of privacy rights and fraud (https://bit.ly/44li2AX).

Opting Out of Algorithmic Traps?

In response to the regulatory requirements on algorithms, some well-known apps have updated their platform policies to enable users to opt out of algorithmic recommendations, pledging to abide by the new algorithmic regulations and consumer data protection regulations. Many introduced this new opt-out option in late 2021 before the algorithmic regulation took effect in 2022. An investigation by *The Paper* (2021) found that 26 out of 28 popular apps introduced ways for users to opt out of personalized recommendations in response to PIPL's stance on the protection of private data. Apart from the opt-out options in recommendation algorithms, Douyin also introduced new features to curb user addiction to the algorithmically-curated feed by inserting five-second pauses along the feed whenever a user had watched for a long time (Qu 2021).

The reality is not always what is indicated or pledged on paper. Xinhua News Agency (2021) has found that platforms have not completely stopped algorithmic price discrimination; rather, their algorithmic traps are more opaque and harder to detect, as expressed in the Chinese term *taolu gengshen le* (套路更深了). Our testing

[63] The 2021 SAMR "Guidance for Internet Platform Classifications (Draft for Comments)" divides Internet platforms into three grades (super, big, and medium-and-small platforms) and six classifications (online sales, services, social networking and entertainment, news and information, financial services, and computing applications). Platform regulations are tiered based on grades and classifications. See the next section for further discussion.

of multiple platforms in early 2022 suggested that they make it difficult for users to find and turn off the recommendation functions; most would only allow a user to turn off advertisements, not content recommendations. Most of the popular platforms automatically turn on the recommendation function again in three to six months after users turn it off. Users would need to go through nine steps every six months on WeChat to turn off advertisement recommendations, or six steps every three months on Meituan and Douyin to do the same. On Weibo, two steps are required to turn off advertisement recommendations every three months. This is not a game for busy bargain hunters who want easy and quick access to products and services at their fingertips.

Chinese experts now ask consumers to be more algorithm-literate. They ask consumers to play with platform algorithms by using multiple platforms to compare prices, or by using different accounts to check prices on the same platforms, to avoid price discrimination. People are asked to game algorithmic traps on digital platforms. On e-commerce and travel booking platforms, for example, time-rich consumers should shop around to take advantage of offers and promotions on various apps. On entertainment platforms, people should train, tame, or confuse algorithms by pretending to like different themes or topics so that they are recommended a more diverse range of topics. Some consumers have opted for digital minimalism, following a global trend (Newport 2019). On Weibo, people started the "Deregistration from JD" hashtag movement as a means of resistance to JD's algorithmic price discrimination in 2021. On Douban, over 30,000 members of the "Anti-Technology-Reliance" group share experiences and tips about digital detoxing (that is, uninstalling apps and limiting social media usage).

In a platform-dependent and cashless society like China, dynamic pricing, cookies, and personalized pricing are all skewed to the interests of digital retailers and platforms. Consumers continue to be subject to algorithmic manipulation, even though they can shop around on different platforms. While opting out of algorithmic traps sounds like a consumer choice, the freedom to opt out or to play with platform algorithms is at best Hobson's choice,[64] circumscribed as it is by the algorithmic ecosystem that underpins the Chinese digital economy.

In sum, China's regulation of algorithmic price discrimination fits into the broader framework of algorithm and AI governance. It is a response to the popular outcry that migrated from smaller online communities to the public sphere via the collective efforts of consumers, scholars, journalists and mainstream media, GONGOs, and government agencies. The process is cumulative and consultative rather than top-down authoritarian. Conciliatory steps have been taken and performative gestures made by big tech companies and the Chinese government in

[64] "Hobson's choice" means having no choice at all.

response to public opinion on regulating algorithmic price discrimination and protecting consumer rights. Such a process of public policymaking illustrates not only the continued effect of scholarship on Chinese authoritarianism through neutral and conciliatory mechanisms (e.g. Heurlin 2016; Deng and Liu 2017), but also shows how China views and manages data, algorithms, and emerging AI technologies in their economic, social, and political governance.

From Regulating Price Discrimination to Governing Algorithms

Algorithmic harm and bias are characteristics of algorithmic capitalism (Mittleman 2022). In the West, cases of hostile, algorithmically-inspired sexism and racism are better known for the harm they inflict on distinct groups of individuals (see Cohn 2019). In everyday life, we mostly encounter seemingly neutral applications of algorithmic systems in non-neutral environments; these have not only perpetuated existing systemic disadvantaging of groups of people, but also created new biases through seemingly scientific and objective algorithms.

As discussed earlier, dynamic pricing discrimination is a type of algorithmic bias that is common practice in businesses like travel booking services, online retail and entertainment services, and mobility (taxi-hailing, bike-rental, truck-hiring) services. Algorithms are commonly applied by digital service platforms to determine the prices that best match the demands of the market and the willingness of consumers to pay. Algorithmic pricing is both efficient and absurd (Thornhill 2022): it is efficient because companies can reach price equilibrium quicker and get bigger profit margins more efficiently in oligopolistic markets; and it is absurd because it is concentrated on expanding the customer base at the expense of customer loyalty. New customers are enticed by lower prices to purchase certain products or services, while loyal customers, loyalty program customers, and premium members are discriminated against by being charged higher prices for the same products or services.

Digital platforms can use algorithms to influence and manipulate information and human behavior regarding political preferences, cultural tastes, and consumer decisions. News recommendation systems use algorithms to control the news content that people consume, in this way constructing and consolidating echo chambers or information cocoons for the masses. E-commerce and entertainment platforms use algorithms that entice people to the wonderland of their preferred products and services, with incentives for them to stay on their platforms and use the services more and for longer. We are all victims of the "algorithmic trap".

Digital platforms use multiple algorithms to trap users/consumers by providing tailor-made products and services based on carefully constructed profiles of individuals and groups of individuals. Huge amounts of data are the oil in the algorithmic engine. The unlimited and unconstrained collection of consumer data by digital platforms is a common practice on Chinese social and digital platforms. Chinese internet companies have all extensively used algorithms—from recommendation algorithms to dynamic pricing algorithms—on their platforms. Chinese regulatory authorities have taken a procedural and granular approach to regulating how data is handled and how algorithms are used through a consultative process with multiple social actors and collaborative efforts among multiple authorities.

Regulating algorithmic price discrimination is one move in the chess game, a means toward an open end rather than an aim. The public outcry was an urban phenomenon joined by middle-class Chinese who were regular consumers of a digital lifestyle that would be considered a luxury by the digitally-left-behind and the rural poor. In the Chinese context, the public complaints were not about the over-collection of personal data by platforms or dynamic pricing in general, but about price discrimination against loyal or VIP customers. Public actors like the media and CCA formed a responsive and interactive network in addressing the issue. The regulatory authorities responded in kind. It seems to be confirmation of the networked authoritarianism argument put forward by MacKinnon (2011).

It must be pointed out, however, that regulating algorithmic price discrimination is not just about consumer protection. Consumer rights may be a safe space in which urban netizens can take action to defend their rights and demand policy changes, but it is also a convenient entry point for the Chinese government to tighten state control of the private sector and ensure national security in the name of protecting consumer rights. This has been illustrated by policy and regulatory rollouts since 2021. One example is the 2022 regulations on recommendation algorithms, jointly issued by nine authorities, which imposed obligations and fines on algorithm-based recommendation service providers, including those providing deep synthesis services.

The move to regulate platform algorithms is part of the ongoing battle to curb the power of transnational big tech companies whose data and key technologies like algorithms are regarded as core to national security. Even before the algorithmic regulation, Chinese regulatory authorities had taken steps to tighten state control of the digital consumer market and to strengthen law enforcement in the digital sector. They acted quickly and decisively in punishing internet companies for violating antitrust and data security laws. Big tech companies like Alibaba, Tencent, and Didi have been repeatedly levied fines for violating anti-monopoly laws (Jolly 2020; Reuters 2021 and 2022; Sohu 2022). While Didi has received most consumer complaints for its algorithmic price discrimination, it is fined for mismanaging drivers' data rather than discriminatory pricing practices.

In other words, regulating algorithmic price discrimination resembles a public performance. The state had already planned to regulate the internet industry;

a series of laws and regulations had been in the making. The public outrage at platforms' algorithmic price discrimination provided the perfect opportunity to roll out new laws and regulations to give more control to the government. Through the official website of Credit China (2019), the state even proposed including digital platforms on the corporate blacklist for violating algorithms regulations. But it is all talk and no action. The state has capitalized on public opinion to achieve policy purposes, that is, reining in the power of private tech companies and sending a signal to the public that the central government is responsive to public opinion.

Platform response to the regulations is also a public performance. Platform companies will push legal-judicial boundaries whenever they can for market and profit maximization. They also collaborate with the Chinese state to achieve the latter's economic, political, and geopolitical aims; as such they enjoy immunity and broad liability in consumer matters. They respond to new legislation through public performances of social responsibility. This is done through joint declarations, public pledges, and self-enforcement of new regulations.

In December 2019, 32 Internet companies met in Xi'an to attend the Chinese Network Trustworthiness Conference organized by the Chinese Federation of Internet Societies. In response to heated public discussion on algorithmic price discrimination on digital platforms, they vowed to build a collaborative mechanism for trust-building in platform economies (Xinhua News Agency 2019). In March 2021, right after the January multi-regulatory authority meeting, ten platforms in food delivery, travel, and online shopping issued a joint pledge to maintain fair online competition, avoid over-collection and misuse of consumer personal information, and stop any price discrimination practices (People Net 2021). The reality, as pointed out earlier, is less rosy than it appears on paper.

The algorithmic regulation exemplifies the public policymaking process, in which a range of stakeholders—platform companies, legal scholars, media, consumer associations, business entities, local legislators, and government authorities—navigate the complex web of interests and negotiate with multiple stakeholders in developing new policies. It is characterized by policy politicization that is socially resonant, ideologically correct, and connects the policy with a more ambitious national political project (Yan and Zhang 2024). Behind the scenes, platform companies are actively involved in lobbying legislators to impose negotiated terms in favor of platforms. In the drafting of China's Internet Security Law and E-Commerce Law, for example, internet companies weighed in during the drafting and consultation processes in an attempt to remove newly-added rules that would increase corporate responsibility and liability (Deng and Liu 2017). For state regulators, appearing tough on big tech companies serves a political purpose: "to gain global legitimacy and reduce internal frictions" (Hong and Xu 2019, 4655).

The 2022 algorithmic regulation, together with the laws and regulations introduced before and after, have largely focused on information and content recommendation algorithms, particularly those that would influence public opinion or prompt social mobilization, as well as protection of gig workers (such as food delivery riders). Consumer rights-related issues like price discrimination on service platforms are secondary concerns among regulators. This reflects China's dual-track and tiered approach to algorithmic governance.

China takes a flexible and iterative approach to regulating algorithms and platforms. Such an approach is tiered to the classification system for digital platforms and technologies. The classification system is based on platform functionality and impact. In functionality, algorithms have five classifications: synthetic and generative algorithms, personalized recommendation algorithms, ranking and selection algorithms, search and filtering algorithms, and scheduling decision algorithms (Recommendation Algorithm Regulations 2021, Article 2). Such a classification is simplistic and does not reflect the reality where any algorithm or set of algorithms could have more than one functionality. Nevertheless, it serves the purpose of assigning regulatory jurisdictions among government agencies. In terms of impact, algorithms are classified into two categories: algorithms for "public opinion or social mobilization" and algorithms for other and commercial purposes (Recommendation Algorithm Regulations 2021, Articles 23 and 24). The latter is subject to the principle of consumer protection, particularly for services that target under-aged and elderly populations. The former category—algorithms for "public opinion or social mobilization"—is under greater scrutiny by regulatory authorities, and is subjected to reporting, security assessment, and human intervention. On news and information platforms (like Toutiao), one may choose to opt out of algorithmic recommendation systems of user-generated content but cannot opt out of system-recommended news items from state media outlets. Algorithmic reproduction of ideology, values, and culture has continued to characterize recommendation platforms (Meng 2021).

As Hong and Xu (2019, 4644) have pointed out when discussing Chinese platform governance, regulation is fragmented, with many moving parts and vectors; it is characterized by the tensions between different policies targeting digital platforms for commerce and for speech. Under such a model, policy silence in some areas is deliberate and tactical, with platform immunity an implicit policy in regulating e-commerce platforms, while strict liability applies to news content platforms. The dual-track, differential governance regime keeps power struggles between public, corporate, and political interests in balance, while engaging in haphazard institutional tweaking and making symbolic regulatory compromises. The provisions in the laws and regulations to govern e-commerce and digital platforms on data security, personal information protection, platform trustworthiness, and algorithmic justice involve minimum participation of consumer rights groups while paying lip service to consumer rights.

The governance of algorithms, therefore, is not just about the governance of algorithms by the state and platforms, but about social governance through algorithms via platforms. To govern through algorithms means balancing the conflicting demands of regulation (that is, of algorithms for "public opinion or social mobilization" and market monopoly) and deregulation (that is, of algorithms for technological and economic development). As Haines (2011, 229) puts it, "regulation is designed to achieve multiple risk reduction goals, from actuarial and sociocultural to political. It is in this complex set of goals that the paradox of regulation arises and where its strengths and limitations can be understood." Regulating algorithmic price discrimination illustrates the process of China's public policymaking and its dual-tracked and tiered approach to algorithms and platform governance.

Conclusion

In this chapter, we have identified key players and events that have stimulated public discussion about algorithmic price discrimination on digital platforms in China. Urban consumers' online murmurings (complaints), along with a few legal cases involving well-known platform service providers, have made "*shashu*" a widely known term and triggered a nationwide discussion among citizen-consumers, state media, GONGOs, and regulators on how to regulate platform algorithms. This chapter has illustrated a consultative and conciliatory, rather than confrontational, process of public policymaking between big tech and regulatory authorities.

Algorithms constitute the organizing force that shapes consumer behavior in marketing and determines individual behavior in social governance. As a new proxy for the power of platform capitalism, algorithms are central to platform governance. While attention has been given to governance *through* algorithms, there are increasing questions about the governance *of* algorithms. Regulating algorithmic price discrimination illustrates the participatory process of Chinese public policymaking (Repnikova and Fang 2018) and the stratified nature of Chinese algorithm governance.

China has taken a differentiated approach to algorithms governance, with varying governing and enforcement strategies for different categories of algorithm via a classification system, that is, a tighter regulatory framework for information and content recommendation algorithms, particularly for those that would influence public opinion and social mobilization, and a looser regulatory framework for service and consumer product algorithms. Algorithmic price discrimination falls in the second category, covering algorithms with lesser social and political impact.

This differentiated approach has been especially obvious since China decided to scale back its "techlash" amid the sluggish economy and decline in international investment in the post-COVID era (The Economist 2022). This explains why the Anti-Monopoly Law is ambiguous in its application of "exploitative abuse" in consumer cases as it is in cases of algorithmic price discrimination on consumer platforms (Han 2022). It also explains the continued use of dynamic pricing algorithms by consumer service platforms after the recommendation algorithm regulations came into force. In the meantime, the Chinese party-state has further tightened its control of public opinion via online networks (Tsai 2016; Zhang 2023) and optimized the employment of platform algorithms to achieve more efficient social governance. From its perspective, the regulation of algorithmic price discrimination is both timely and politically convenient.

As Xin Dai (chapter 2 of this volume) has illustrated, algorithmic price discrimination is a kind of ADM failure that has now faded away from its high-noise phase and become less salient in public discourse. The term "*shashu*" has not become a historical practice despite new laws and regulations on data security, personal privacy, and algorithmic regulations. Trapped by platform algorithms, Chinese consumers have no choice but succumb to the banality of algorithmic manipulation and surveillance.

References

Bai, Ying, Wenjuan Zhao, Cheng. Shen, and Shao Luo. 2018. "Bimian Dashuju 'Shashu', Guifan Dashou Xingwei: Jujiao Dianshangfa Cao'an Sanshen Redian" [Avoid price discrimination and regulate tie-in sales: A focus analysis of hot topics of the 3rd draft of the E-commerce law]. *Xinhua Net*, June 20, 2018. http://www.npc.gov.cn/zgrdw/npc/cwhhy/13jcwh/2018-06/20/content_2056254.htm.

CAC (Cyberspace Administration of China), MIIT (Ministry of Industry and Information Technology), MPS (Ministry of Public Security), and SAMR (State Administration for Market Regulation). 2021. "Recommendation Algorithm Regulations." gov.cn. December 31, 2021. https://www.gov.cn/zhengce/zhengceku/2022-01/04/content_5666429.htm.

CCA (China Consumers Association). 2021. "Zhongxiaoxie Zai Jing Zhaokai 'Wangluoxiaofei Lingyu Suanfa Guizhi Yu Xiaofeizhe Baohu Zuotanhui'" [CCA convened a meeting over 'regulations of algorithms in the e-commerce sector and consumer protection']. CCA. January 7, 2021. https://www.cca.org.cn/zxsd/detail/29896.html.

Chow, Eugene K. 2018. "Why China Keeps Falling for Pyramid Schemes: Pyramid Schemes Are Big Business in China—Much to the Government's Chagrin." *The Diplomat*, March 5, 2018. https://thediplomat.com/2018/03/why-china-keeps-falling-for-pyramid-schemes/.

Classic Case. 2021. "Dashuju Shashu, Xiecheng Bei Pan Tui Yi Pei San" 大数据杀熟, 携程被判退一赔三 [Price discrimination confirmed, Ctrip ordered to return consumer payment and pay triple-time compensation]. Sohu. July 26, 2021. https://www.sohu.com/a/479703386_121106854.

Cohn, Jonathan. 2019. "Google's Algorithms Discriminate against Women and People of Colour." *The Conversation*, April 25, 2019. https://theconversation.com/googles-algorithms-discriminate-against-women-and-people-of-colour-112516.

Credit China. 2019. "Gei Dashuju Shashu Daishang Falv Peitou" 给大数据"杀熟"戴上法律辔头 [Bridle the shashu price discrimination with laws]. Credit China. April 2, 2019. https://www.creditchina.gov.cn/home/zhuantizhuanlan/fengxiantishi/xinyongfengxiandajiatan/201904/t20190402_151581.html.

Creemers, Rogier, Graham Webster, and Helen Toner. 2022. "Translation: Internet Information Service Algorithmic Recommendation Management Provisions—Effective March 1, 2022." *DigiChina*, January 10, 2022. https://digichina.stanford.edu/work/translation-internet-information-service-algorithmic-recommendation-management-provisions-effective-march-1-2022/.

Danaher, John, Michael J. Hogan, Chris Noone, Rónán Kennedy, Anthony Behan, Aisling De Paor, Heike Felzmann, Muki Haklay, Su-Ming Khoo, John Morison, Maria Helen Murphy, Niall O'Brolchain, Burkhard Schafer, and Kalpana Shankar. 2017. "Algorithmic Governance: Developing a Research Agenda through the Power of Collective Intelligence." *Big Data & Society* 4 (2) (July–December): 1–21. https://doi.org/10.1177/2053951717726554.

Deng, Jinting, and Pinxin Liu. 2017. "Consultative Authoritarianism: The Drafting of China's Internet Security Law and E-Commerce Law." *Journal of Contemporary China* 26 (107): 679–695. https://doi.org/10.1080/10670564.2017.1305488.

Dong, Xue, Siqi Cheng, Lin Zhou, and Zhengdong Wu. 2021. "Jizhe Shice Faxian: Xinzheng Fabu Hou, Yixie Hulianwang Pingtai Shashu Bu Gai Qie Taolu Gengshen" [Journalists' fieldwork found that some internet platforms continued to adopt shashu and their strategies were more covert than before even after the promulgation of new anti-trust policies]. *Xinhua Net*, March 15, 2021. http://www.xinhuanet.com/fortune/2021-03/15/c_1127213756.htm.

Du, Yuanhun, Ruobai Zhang, and Xingyi Qu. 2018. "51.3% Xiaofeizhe Zaoyu Guo Dashuju Shashu" [51.3% of consumers encountered big-data algorithmic price discrimination]. *China Youth*, March 15, 2018. http://zqb.cyol.com/html/2018-03/15/nw.D110000zgqnb_20180315_1-07.htm.

Feng, Qiuyu, Dan Zhang, and Yuan Li. 2021. "Dashujushashu Heshixiu? Zhuanjia: Xianshang 'Jiageqishi' Gengyinbi" [When can price discrimination practices be stopped? Experts: Online 'price discrimination' is harder to find out]. *Guangzhou Daily*, August 18, 2021. http://www.chinanews.com/cj/2021/08-18/9545874.shtml.

Fingar, Thomas. 2016. "Government China Specialists: Scholar Officials and Official Scholars." In *American Studies of Contemporary China*, edited by David L. Shambaugh, 176–195. London and New York: Routledge. https://doi.org/10.4324/9781315484570-12.

Gugugu. 2018. "Dashuju Zhixia, Ni Jiushi Bei Shashu De Yuandatou" [Under the rule of big data, you will be played for a sucker!]. *Vista Tianxia Jiaodian*, April 27, 2018. https://www.163.com/dy/article/DGE8RSSB0512DALL.html.

Haines, Fiona. 2011. *The Paradox of Regulation: What Regulation Can Achieve and What It Cannot.* Cheltenham, UK: Edward Elgar.

Han, Wei. 2022. "The New Anti-Monopoly Law of the People's Republic of China: Changes and Remaining Issues." August 7, 2022. https://www.pymnts.com/cpi_posts/the-new-anti-monopoly-law-of-the-peoples-republic-of-china-changes-and-remaining-issues/.

Heurlin, Christopher. 2016. *Responsive Authoritarianism in China: Land, Protests, and Policy Making.* Cambridge: Cambridge University Press. https://doi.org/10.1017/CBO9781316443019.

Hong, Yu, and Jian Xu. 2019. "Toward Fragmented Platform Governance in China: Through the Lens of Alibaba and the Legal-Judicial System." *International Journal of Communication* 13: 4642–4662.

Huld, Arendse. 2022. "What Has Changed in China's Amended Anti-Monopoly Law?" *China Briefing*, July 11, 2022. https://www.china-briefing.com/news/what-has-changed-in-chinas-amended-anti-monopoly-law/.

Jin, Xin. 2021. "Geren Xinxi Baohufa Cao'an Jinru Ershen, Qianghua Hulianwang Pingtai Geren Xinxi Baohu Yiwu" [The draft of Personal Information Protection Law is under the 2nd review: the revision attempts to reinforce platforms' personal information protection obligations]. *People's Daily*, April 27, 2021. http://www.npc.gov.cn/npc/c2/c30834/202104/t20210427_311217.html

Jolly, Jasper. 2020. "Ant Group Forced to Suspend Biggest Share Offering in History." *The Guardian*, November 4, 2020. https://www.theguardian.com/business/2020/nov/03/biggest-share-offering-in-history-on-hold-as-ant-group-suspends-launch.

Katzenbach, Christian, and Lena Ulbricht. 2019. "Algorithmic Governance." *Internet Policy Review* 8 (4). https://doi.org/10.14763/2019.4.1424.

Li, Ling. 2021a. "Shashu, Lanyong Shuju Nan Jianguang? Renda Daibiao Yang Song: Zengjia Fan Longduanfa Guizhi Tiaokuan" 杀熟、滥用数据难监管?人大代表杨松:增加反垄断法规制条款 [Hard to supervise and control shashu and misuse of data? People's representative Yang Song: add binding articles in anti-monopoly laws]. *Southern Metropolis Daily*, March 13, 2021. https://m.mp.oeeee.com/a/BAAFRD000020210313453350.html.

Li, Ling. 2021b. "Huiyuan ding jiudian gui yibei, yonghu gaoying xiecheng! zheshi Dashujushashu diyian ma [VIP members paying double the price, Ctrip use on the case against Ctrip! Is this the first shashu case?]." *Southern Metropolis Daily*, July 16, 2021. https://m.mp.oeeee.com/a/BAAFRD000020210716521350.html

Lobato, Ramon. 2021. "Free, bundled or personalized? Rethinking price and value in digital distribution." In *Digital Media Distribution: Portals, Platforms, Pipelines*, edited by Paul McDonald, Courtney Brannon Donoghue, and Timothy Havens, 314–334. NYU Press.

Lu, Yue. 2021. "Yichang xinxi baohu de boyi 一场信息保护的博弈 [A game of information protection]." *Xinhua Net*, October 20, 2021. http://www.xinhuanet.com/legal/2021-10/20/c_1127974932.htm

MacKinnon, Rebecca. 2011. "Liberation Technology: China's 'Networked Authoritarianism'." *Journal of Democracy* 22 (2): 32–46. https://doi.org/10.1353/jod.2011.0033.

Meng, Jing. 2021. "Discursive Contestations of Algorithms: A Case Study of Recommendation Platforms in China." *Chinese Journal of Communication* 14 (3): 313–328. https://doi.org/10.1080/17544750.2021.1875491.

Mittelman, James H. 2022. "The Power of Algorithmic Capitalism." *International Critical Thought* 12 (3): 448–469. https://doi.org/10.1080/21598282.2022.2070858.

Moor, Liz, and Celia Lury. 2018. "Price and the Person: Markets, Discrimination, and Personhood." *Journal of Cultural Economy* 11 (6): 501–513.

Newport, Cal. 2019. *Digital Minimalism: Choosing a Focused Life in a Noisy World*. Portfolio.

PDSM. 2021. "Dache ruanjian de mimi, kan zhe yipian jiu goule [The secret of taxi ride-hailing software, just read this article]." Sohu. March 02, 2021. https://www.sohu.com/a/453676394_184714

People Net. 2020. "Dashujushashu Zao Goubing, Jianguan Lifa Jidai Wanshan" [Price discrimination has been widely criticized: It is urgent to improve relevant regulations and laws]. 163.com. March 16, 2020. https://www.163.com/dy/article/F7R8SDLQ05346936.html

PeopleNet. 2021. "Vipshop, JD, Meituan deng 10 jia hulianwang pingtai chengnuo: buliyong dashuju shashu 唯品会、京东、美团等10家互联网平台承诺:不利用大数据'杀熟' [10 internet platforms including Vipshop, JD, and Meituan promise not to use shashu]." *PeopleNet*, April 9, 2021. http://finance.people.com.cn/n1/2021/0409/c1004-32074088.html

Qu, Tracy. 2021. "TikTok's China Sibling Douyin Launches Mandatory Five-Second Pauses in Video Feed to Curb User Addiction." *SCMP*, October 22, 2021. https://www.scmp.com/tech/policy/article/3153292/tiktoks-china-sibling-douyin-launches-mandatory-five-second-pauses

Repnikova, Maria, and Kecheng Fang. 2018. "Authoritarian Participatory Persuasion 2.0: Netizens as Thought Work Collaborators in China." *Journal of Contemporary China* 27 (113): 763–779. https://doi.org/10.1080/10670564.2018.1458063.

Reuters. 2022. "China's Didi Global fined $1.7 billion for 'serious' personal information protection violations." *ABC News*, July 21, 2022. https://www.abc.net.au/news/2022-07-21/didi-fined-billion-dollars-for-data-protection-breaches/101259554

Reuters. 2021. "Didi App Suspended in China over Data Protection." *CNBC*, July 4, 2021. https://www.cnbc.com/2021/07/04/didi-app-suspended-in-china-over-data-protection.html

Shao, Lanjie. 2019. "Beijing Xiaoxie: Wanggou, Zaixian Lvyou, Wangyueche Dashujushashu Wenti Zuiduo [Beijing Consumers' Association: Big data-enabled price discrimination is the most serious in the online shopping sector, the online travel service sector, and the taxi ride-hailing service sector]." *Beijing Business Daily*, March 27, 2019. https://tech.sina.com.cn/i/2019-03-27/doc-ihsxncvh5927307.shtml

SAMR. 2021. "Shichangjianguanzongju, zhongyangwangxinban, shuiwuzongju lianhe zhaokai hulianwang pingtai qiye xingzheng zhidaohui [SAMR, CAC, and STA convened a meeting guiding internet platforms]." SAMR. April 13, 2021. https://www.samr.gov.cn/xw/zj/202104/t20210413_327785.html

Sohu. 2022. "Ali, tengxun weifan fan longduanfa, bei fa 850 wan! Weihe laoshi beifa? 阿里、腾讯违反 反垄断法 ,被罚850万 为何老是被罚?[Alibaba and Tencent are fined 8.5 million for violating the anti-monopoly laws. Why are they always fined?]." Sohu. July 10, 2022. https://www.sohu.com/a/565974387_120524731

The Economist. 2022. "Is This the Beginning of the End of China's Techlash?" *The Economist*, March 19, 2022. https://www.economist.com/business/is-this-the-beginning-of-the-end-of-chinas-techlash/21808208

The Paper. 2020. "456 tiao tousu hou, dashuju shashu neng you duo meipu 456条投诉背后,大数据杀熟能有多没谱 [Behind the 456 complaints: how ridiculous is shashu]." *The Paper*, October 21, 2020. https://m.thepaper.cn/kuaibao_detail.jsp?contid=9399206&from=kuaibao

The Paper. 2021. "APP buneng qiangzhi gexinghua tuijian le, ni zhidao ruhe guanbi ma? APP不能强制个性化推荐了,你知道如何关闭吗?[Do you know how to turn off personalized recommendations on apps?]." *The Paper*, August 25, 2021. https://m.thepaper.cn/newsDetail_forward_14203101

Thornhill, John. 2022. "Algorithmic Pricing Is Both Efficient and Absurd." *Financial Times*, May 19, 2022. https://www.ft.com/content/1fb2792a-c4e0-41b5-83bb-9e1912aa5275

Tsai, Wen-Hsuan. 2016. "How 'Networked Authoritarianism' Was Operationalized in China: Methods and Procedures of Public Opinion Control." *Journal of Contemporary China* 25 (101): 731–744. https://doi.org/10.1080/10670564.2016.1160506.

Ulbricht, Lena, and Karen Yeung. 2022. "Algorithmic Regulation: A Maturing Concept for Investigating Regulation of and through Algorithms." *Regulation & Governance* 16 (1): 3–22. https://doi.org/10.1111/rego.12437.

Wang, Wei, and Tingting Zhao. 2019. "Beijingshi xiaoxie fabu 'dashujushashu' wenti diaocha jieguo [Beijing Consumers' Association published the survey outcome over price discrimination]." *Beijing Youth*, March 28, 2019. https://baijiahao.baidu.com/s?id=1629203913661292412&wfr=spider&for=pc

Wu, Jialing. 2020. "Pingtai Lv Xian Dashujushashu Zhengyi Beihou: Huo Sheji Longduan Xingwei, Xiaofeizhe Juzheng Nan [Behind platforms' repeated shashu scandals: It may involve monopolistic behavior and consumers face the burden of proof]." *Southern Metropolis Daily*, December 19, 2020. https://m.mp.oeeee.com/a/BAAFRD000020201219397613.html

Wang, Wanying. 2019. "Beijingshi xiaoxie fabu 'shashu' diaocha, wanggou, zaixian lvyou he wangyueche wenti zuiduo [Beijing Consumers' Association publishes price discrimination survey, problems are the most serious in the online shopping sector, the online travel service sector, and the taxi ride-hailing service sector]." *Beijing Evening News*, March 27, 2019. http://www.ce.cn/cysc/tech/gd2012/201903/27/t20190327_31751908.shtml

Wang, Wei, and Tingting Zhao. 2019. "Beijing shi xiaoxie fabu dashuju shashu wenti diaocha jieguo 北京市消协发布'大数据杀熟'问题调查结果 [Beijing Consumers' Association releases the survey result of big-data shashu]." *PeopleNet*, March 28, 2019. http://finance.people.com.cn/n1/2019/0328/c1004-30999903.html

Xinhua News Agency. 2019. "32 jia hulianwang danwei gongtong qidong pingtai jingji linglv xinyong jianshe hezuo jizhi 32家互联网单位共同启动平台经济领域信用建设合作机制 [32 internet companies initiated a mechanism for credit construction in platform economy]." gov.cn. December 2, 2019. https://www.gov.cn/xinwen/2019-12/02/content_5457631.htm

Xinhua News Agency. 2021. "Jizhe shice faxian: xinzheng fabu hou, yixie hulianwang pingtai shashu bugai qie taolu geng sheng 记者实测发现：新政发布后，一些互联网平台'杀熟'不改且套路更深 [Exposed: after the new policy was announced, some online platforms have not changed their shashu practices; instead, there are more traps]." *xinhua.net*, March 15, 2021. http://www.xinhuanet.com/fortune/2021-03/15/c_1127213756.htm

Yan, Xiaojun, La Li, and Zhengyu Zhang. 2024. "Politicization as a Policy Instrument: China's Politicized Policy Narrative of Environmental Protection and Control of Its Social Resonance." *Journal of Contemporary China* 33 (149): 755–773.

Yu, Mengmeng. 2019. "Xiecheng Bei Zhi 'Dashujushashu', Hulianwang Qiye 'Shashu' Sanzhong Taolu, Ni Zhong Guo Zhao Ma [Xiecheng is accused of involving in shashu, have encountered the three tricks played by digital companies]." *Zhongguo Zhengquan Bao*, March 12, 2019. http://www.nbd.com.cn/articles/2019-03-12/1309227.html

Zhai, Dongdong. 2018. "Wangluo Gongsi Yong Dashujushashu: Jiudian Fangjia 300 Yuan bian 380 Yuan [IT companies profit from loyal customers by using big data technologies: Hotel room price raises from ¥300 to ¥380]." *Science & Technology Daily*, February 28, 2018. http://finance.sina.com.cn/consume/puguangtai/2018-02-28/doc-ifyrwsqk0428812.shtml

Zhang, Albert. 2023. "Gaming Public Opinion: The CCP's Increasingly Sophisticated Cyber-Enabled Influence Operations." *ASPI Policy Brief Report* 71. https://www.aspi.org.au/report/gaming-public-opinion

Zhang, Liqiao. 2021. "Beiyiwangquan Zhidu Kuangjia Ji Yinru Zhongguo De Kexingxing [The Regulatory Frame of the Right to Be Forgotten and the Possibility of Introducing It to China]." *Internet Finance & Law*. https://www.finlaw.pku.edu.cn/hlwjryfl/xk_hljryfl/239436.htm

Zhang, Mengjiao. 2020. "Zhuanjia Jiexi 'Dashujushashu': Xiaofeizhe Weiquan Ke Tiqi Fanlongduan Susong [Experts Explain BDEPD: Consumers Could File Anti-Trust Lawsuits to Protect Their Rights]." *Justice Net*, December 31, 2020. http://news.jcrb.com/jsxw/2020/202012/t20201231_2238123.html

Zhang, Tianpei. 2021. "Geren Xinxi Baohufa Cao'an Jinru Sanshen, Jianquan Wanshan Geren Xinxi Baohu Zhidu Guize [The draft of personal information protection law is under the 3rd review: the revision attempts to ameliorate personal information protection regulations]." *People's Daily*, April 27, 2021. http://www.npc.gov.cn/npc/c2/c30834/202108/t20210818_312929.html

Zhao, Peng. 2019. "Jin sancheng yonghu tousu, dashujushashu gai shui juzheng [Only 30% of victims complained, who should be the one providing evidence in big data-enabled price discrimination cases]." *Beijing Daily*, March 28, 2019. http://media.people.com.cn/n1/2019/0328/c40606-30999365.html

Zhao, Yiwei. 2018. "Boyi Dianshangfa: Weirao Fa'an Xize De Zhengyi Yuanwei Pingxi [Wrestles in making the E-commerce law: debates around specific articles are far from over]." *China News Weekly*, September 9, 2018. https://www.chinanews.com.cn/m/sh/2018/09-09/8622393.shtml

Zhao, Yue. 2019. "Beijing Xiaoxie Fabu Dashujushashu Diaocha Jieguo [Beijing Consumers' Association publishes the result of the survey on big data-enabled price discrimination]." *People's Daily*, April 10, 2019. https://smart.huanqiu.com/article/9CaKrnKjF3K

CHAPTER 10

The Algorithmic Divide in China and An Emerging Comparative Research Agenda

Peter K. Yu

Abstract

Recognizing that past scholarship on the digital divide can provide helpful insights into research on the algorithmic divide, this chapter begins by identifying the similarities and differences between these two inequitable gaps. The chapter then discusses the importance of studying the algorithmic divide in China and how this study can build on, illuminate, and create synergy with China-related academic and policy research in other areas. To highlight the potential comparative insights provided by studying the algorithmic divide, this chapter concludes by examining three sets of policy responses that commentators have proposed in legal and policy literature to bridge this divide. It further contextualizes these responses in relation to local conditions in China.

Keywords: Algorithmic divide; Artificial intelligence; Automated decision-making; Big data analytics; Digital divide; Machine learning

Introduction

With the arrival of big data analytics, machine learning, and artificial intelligence (AI), governments at both the national and subnational levels have been eager to deploy automated decision-making (ADM) systems to detect and recognize patterns, predict and shape preferences, and ultimately streamline and improve governance. Various chapters in this volume have explored the myriad opportunities and challenges provided by the increased use of algorithms and ADM systems as well as the emergence of what commentators have called an "algorithmic society" (Balkin 2017, 1219).

One topic that has been underexplored in ADM literature is the gap between those who have access to, or proficiency in, algorithmically enhanced or AI-driven technological products and services and those who do not (Goggin and Soldatić 2022). This proverbial gap resembles the digital divide—the widely documented disparities between those who have access to the Internet, information and communication technology (ICT), and digital content and those who do not. For the past three decades, scholars in communication studies and other disciplines have conducted extensive research on the digital divide and in relation to digital inclusion and equality (Compaine 2001; Hargittai 2021; Mack 2001; Norris 2001; Peacock 2019; Ragnedda 2017; Ragnedda and Muschert 2013, 2018; U.N. High-Level Panel 2019; Van Dijk 2005, 2020; Warschauer 2003; Yu 2002). When the digital divide is linked to the wide societal use of algorithms and ADM systems, it slowly morphs into what other commentators and I have called the "algorithmic divide" (Yu 2020c). Some commentators also use the terms the "new digital divide" and the "AI divide" (Bentley et al. 2024, 903; Jarrett 2017; Susarla 2019). The latter recognizes that the algorithms used in ADM systems often involve machine learning and AI.

Today, the newly emerged and fast-expanding algorithmic divide affects the entire world and deserves urgent scholarly and policy attention. As the U.N. Secretary-General's High-Level Advisory Body on Artificial Intelligence (U.N. High-Level Advisory Body) (2024, 7) recently declares in its final report:

> Left ungoverned, ... AI's opportunities may not manifest or be distributed equitably. Widening digital divides could limit the benefits of AI to a handful of States, companies and individuals. Missed uses—failing to take advantage of and share AI-related benefits because of lack of trust or missing enablers such as capacity gaps and ineffective governance—could limit the opportunity envelope.

China is no exception. Just as we need to pay greater attention to automated governance and the use of algorithms and ADM systems in the country, we can also benefit from a deeper appreciation and understanding of the algorithmic divide in China. For instance, it will be worthwhile interrogating whether this divide should be examined the same way in China as in other countries. Equally important is the question of whether China will provide a useful multidisciplinary case study on the algorithmic divide. Will this case study inform analyses of the algorithmic divide in other parts of the world or at the global level? Within China, will this study offer valuable insights into Chinese technology law, policy, and regulation?

A good starting point for studying the algorithmic divide in China is the historical studies on the country's digital divide—whether based on geography, demographics, socioeconomic conditions, or other factors. However, these studies remain scant, especially those published in English (CNNIC 2024, 18–20; Pick and Sarkar 2015, 113–54; Song et al. 2020). Thus, anybody eager to study the algorithmic divide in China will face greater challenges than those studying the same divide in other parts of the world. In view of these challenges, it is worth examining how researchers should approach this underexplored topic. Do the digital and algorithmic divides provide useful parallels? Can scholarship on either divide in China strengthen and benefit from China-related academic and policy research in other areas? What comparative insights can one glean from studying the algorithmic divide in China?

Recognizing that past scholarship on the digital divide can provide helpful insights into research on the algorithmic divide, this chapter begins by identifying the similarities and differences between these two inequitable gaps. This chapter then discusses the importance of studying the algorithmic divide in China and how this study can build on, illuminate, and create synergy with China-related academic and policy research in other areas. To highlight the potential comparative insights provided by studying the algorithmic divide, this chapter concludes by examining three sets of policy responses that commentators have proposed in legal and policy literature to bridge this divide. It further contextualizes these responses in relation to local conditions in China. It is my hope that by outlining many possible directions for research on the algorithmic divide in China, this chapter will encourage academic and policy researchers to devote greater attention to this underexplored topic.

From the Digital Divide to the Algorithmic Divide

Academic and policy research and media reports surrounding the digital divide began to emerge in the mid-1990s, when the use of the Internet started to enter the mainstream. From 1995 to 2000, the Clinton Administration in the United States released four detailed surveys in the series *Falling through the Net* (National Telecommunications and Information Administration 1995, 1998, 1999, 2000). After the launch of this series, books and journal articles covering the digital divide began to appear (Compaine 2001; Mack 2001; Norris 2001; Warschauer 2003; Yu 2002).

While the digital divide no longer attracts the same amount of attention as it used to, it remains relevant in public policy debates and has returned from time to time, especially in the run-up to political elections. The topic of digital inclusion has also garnered interest at the United Nations, such as in the final reports of both the U.N. High-Level Advisory Body (2024) and the United Nations Secretary-General's High-Level Panel on Digital Cooperation (U.N. High-Level Panel) (2019). Target 9.C of the U.N. Sustainable Development Goals explicitly calls for efforts to "[s]ignificantly increase access to information and communications technology and strive to provide universal and affordable access to the Internet in least developed countries by 2020." During the COVID-19 pandemic, U.N. Secretary-General António Guterres went even further in calling the digital divide "a matter of life and death for people who [we]re unable to access essential health-care information" (United Nations 2020).

In the past few years, commentators have begun paying greater attention to the algorithmic divide. Given the insights scholarship on the digital divide can provide into research on the algorithmic divide, this section draws on the author's prior research on the former to identify five attributes that can be used to systematically analyze the latter: (1) awareness; (2) access; (3) affordability; (4) availability; and (5) adaptability (Yu 2002, 8–16). Although the discussion focuses on the similarities between these two inequitable gaps, it also highlights many notable differences. Readers are therefore invited to draw their own conclusions as to whether the digital and algorithmic divides provide useful parallels, as well as what research approaches and methodologies would be the most appropriate and optimal for studying the algorithmic divide in China.

Because the algorithmic divide has impacted the entire world, not just China, this section provides illustrations using geographically neutral examples. The next section, by contrast, will focus more specifically on China. At the outset, it is worth pointing out that some readers may disagree with my classification of the less inclusive side of the algorithmic divide as unfortunate, due to the many challenges and harm associated with greater use of big data analytics, machine learning, and AI. Despite the potential silver linings, this chapter uses the word "unfortunate" to highlight the many missed political, social, economic, cultural, educational, and career opportunities provided by these new technologies.

Awareness

Those on the unfortunate side of the digital divide can easily notice their being left out of the Internet revolution (Louie 1999), especially after the Internet entered the mainstream in the mid-1990s. By contrast, those on the unfortunate side of the algorithmic divide may have greater difficulty discovering their exclusion from machine learning and AI (Frischmann and Selinger 2018, 13). Indeed, many individuals on this side may not appreciate how the increased use of machine-learning algorithms and ADM systems can impact their lives—both positively and negatively. Even among those taking note of these impacts, most will have a very limited understanding of how algorithms operate (Chander 2017, 1040; Domingos 2012, 5; Eubanks 2017, 184–85; Kroll et al. 2017, 638; Rainie and Anderson 2017, 19).

In an algorithmic society, individuals will need to become more aware of the strengths and weaknesses of algorithmically enhanced or AI-driven technological products and services, including in the contexts of ADM and automated governance. While these new products and services enable individuals to complete tasks that they otherwise could not accomplish with traditional computing technology, these products and services could create harm in the form of biases, bugs, and dehumanization (K. Lee 2018, 173; Michaels 2020, 1103; Noto La Diega 2018, 33; Rainie and Anderson 2017, 9–11, 42–56). Frank Pasquale (2015) laments how we now live in a "black box society," while Cathy O'Neil (2016) refers to machine-learning algorithms as "weapons of math destruction."

Access

Access is the most widely discussed attribute of the digital divide. The use of algorithms can provide important individual and societal benefits—and could even help bridge the digital divide—yet not everybody has access to algorithmically enhanced or AI-driven technological products and services.

At the domestic level, individuals will be shut out when they cannot afford these products and services, cannot find them on the local market, or do not possess the needed skills to use them, or use them effectively (Yu 2020c, 341–43). At the global level—whether in developing countries or the less developed parts of emerging economies—this access barrier will become even more substantial (De-Arteaga et al. 2018, 2; Yu 2002, 4–5). As the U.N. High-Level Panel (2019, 6) reminds us: "Well more than half the world's population still either lacks affordable access to the internet or is using only a fraction of its potential despite being connected."

To a large extent, much of the prior research on ICT for development can provide instructive lessons for addressing development-related challenges in the age of AI and in the context of an algorithmic society (De-Arteaga et al. 2018). Among the strategies proposed for developing countries are an increased ability to handle small and messy datasets, the development of intelligent data-acquisition strategies and compression algorithms, the creation of transfer-learning models for low-resource languages, the facilitation of machine learning with limited computational capabilities, and the utilization of decision support systems (De-Arteaga et al. 2018, 9–10).

Affordability

Affordability goes hand in hand with access and is often lumped together under a single umbrella as "affordable access," but the two attributes involve different considerations and are better analyzed separately.

While the lack of economic and technological resources may lead to inaccessibility, it can also determine the type of product and service that an individual can access and the frequency at which that individual can utilize the selected product or service. Because affordability also limits one's ability to "upgrad[e] ... equipment, software, and training support" (Yu 2002, 12), this attribute may further affect the overall product or service quality.

To a large extent, affordability determines not only individual access to machine learning and AI but also one's ability to fully participate in an algorithmic society. The less access one can afford, the fewer benefits one will be able to derive from algorithmically enhanced or AI-driven technological products and services, and the less likely one is to fully realize the promise of machine learning and AI.

Availability

There is a general assumption that individuals will have the needed technological products or services if machine-learning capabilities become accessible and affordable (the first two attributes). Yet, this assumption does not always hold given the differing individual needs for products and services (Yu 2002, 13). Indeed, it is not uncommon to find that the specific type of product or service needed by an individual does not exist.

Even if that product or service exists, it may feature algorithms designed by those who do not fully grasp the user's specific needs and conditions, especially when the user resides in the developing world. As Ralph Hamann (2018) laments: "AI algorithms are developed almost entirely in developed regions. Thus they may not sufficiently reflect the contexts and priorities of developing countries." To improve algorithms, commentators have repeatedly called for greater inclusivity in both the design process and the data used for training algorithms (Rainie and Anderson 2017, 23; U.N. High-Level Panel 2019, 29–30; Yu 2021, 280–84).

Since the mid-2010s, commentators have widely discussed the problem of algorithmic bias and discrimination (Barocas and Selbst 2016; Bornstein 2018; Huq 2019; Kim 2017b; Packin and Lev-Aretz 2018; Selbst 2017; Yu 2020c, 354–59; Zarsky 2014). While this problem has produced undesirable outcomes harming select individuals, it could shut these individuals out of access entirely. Whether intentional or not, algorithmic bias and discrimination threaten to take away the benefits provided by machine learning and AI to a large segment of the population.

Adaptability

If individuals are to succeed in an algorithmic society, they will need to take advantage of the different algorithmically enhanced or AI-driven technological products and services. They will also need to adapt these new tools to their own needs (Yu 2002, 15). Without adaptation, these tools may not enable these individuals to fully realize the promise of machine learning and AI.

Adaptability, however, requires both knowledge and understanding. In an algorithmic society, literacy is just as important as awareness. To enhance algorithmic literacy, policymakers will need to put in place institutions, programs, mechanisms, and other measures, which the last section of this chapter will further discuss. Should those on the unfortunate side of the algorithmic divide fail to adequately adapt to this fast-changing technological environment, they will likely be left behind (Rainie and Anderson 2017, 63).

China as a Multidisciplinary Case Study

Although policymakers, commentators, and the mass media have only just begun to pay attention to the algorithmic divide, and it remains difficult to locate China-related research on this topic, this section will explain why studying this divide in China is important and highly beneficial. This section will further argue that this case study can build on, illuminate, and create synergy with China-related academic and policy research in other areas.

The Digital and Algorithmic Divides in China

China has been noted for its wide subnational disparities in access to the Internet, ICT, and digital content (Yu 2012, 395–96; 2014, 98–99). As the authors of a recent study recount:

> The spatial distribution of digital divide [in China] varied among ... three levels. For the first order of digital divide [which focuses on access], the cities with the highest ICT access value in 2016 were Guangzhou, Shenzhen, Hangzhou, Suzhou and Shanghai with more than 90 [in the Digital Divide Index (DDI)], with the lowest value of about 17 in Liangshan and Ganzi Autonomous Prefectures in Sichuan Province, Wenshan Autonomous Prefecture in Yunnan Province, and Neijiang in Guizhou Province .... While the second order digital divide [which emphasizes skills] shows similar spatial patterns ... with the highest DDI values in prefectures with metropolises [and] metropolitan areas, including Shenzhen, Shanghai, Hangzhou, Beijing, Guangzhou, Suzhou, and Ningbo. The lowest DDI values are in rural prefectures in the northwest and southwest China, including Longnan and Gannan in Gansu Provinces, Bazhong and Liangshan Autonomous Prefectures in Sichuan Province, Guoluo Autonomous Prefectures in Qinghai Province, and Wenshan Autonomous Prefectures in Yunnan Province. As for the third order digital divide [which involves ICT outcomes], ... there are significant differences ... among cities, and ... spatial clustering in certain regions. For example, most high level cities are concentrated in coastal areas, such as Shenzhen, Guangzhou, Zhongshan, Zhuhai, and Dongguan in the Pearl River Delta, and Hangzhou, Jinhua, Jiaxing, Shanghai and Ningbo in the Yangtze River Delta. (Song et al. 2020, 6)

This study is consistent with the statistical surveys on Internet developments in China that the state-run China Internet Network Information Center (CNNIC) has released biannually since December 1997. Using data up to June 2024, the CNNIC's November 2024 report states that China had nearly 1.1 billion netizens, with an Internet penetration rate of 85.3% in urban areas and 63.8% in rural areas (CNNIC

2024, 13, 16). That report put "the number of non-Internet users in China [at] 310 million" (CNNIC 2024, 18).

Although the discussion of the digital divide in China has thus far focused on geographical disparities, this divide also features disparities based on demographics (in particular, gender and age groups) and other factors (such as educational backgrounds, income levels, and technological competence). As the CNNIC (2024, 19) continues to observe in its report:

> The main reasons that non-Internet users do not go online were a lack of skills, educational limitations, insufficient devices, and age. 49.0% of non-Internet users did not go online, because they did not know how to use computers/networks; 27.6% of them due to not mastering Pinyin or other literacy limitations; 19.0% of them due to not having the necessary devices, like a computer; and 15.3% of them due to being too young/old.

It is therefore no surprise that commentators have considered these factors increasingly more important than geography.

To the extent that Internet access—whether through traditional computers, mobile phones, or other digital devices—is the prerequisite for active participation in today's algorithmic society, the disparities identified in the CNNIC report and the earlier study provides helpful insights into the extent of the algorithmic divide in China. While most netizens in the country now have easy access to smartphones, laptops, and similar devices, their use and understanding of ICT vary significantly (Bentley et al. 2024, 902–03). Instead of using ICT as a political, socioeconomic, cultural, educational, or career-oriented tool, many netizens may use it for "gaming, social media, and [other] leisure activities" (Bentley et al. 2024, 903). The disparities in the use and understanding of this technology became particularly salient during the COVID-19 pandemic, when individuals and their households needed technology to obtain health-related information and when state authorities actively deployed technological measures to facilitate contact tracing, quarantines, virus testing, and vaccination (Shao and Kostka 2023, 4; Yang 2022, 54; H. Yu 2024, 3).

Because the previous section already identified the similarities and differences between the digital and algorithmic divides and the current discussion provides information about the Chinese context, the remainder of this section turns to the potential connections between scholarship on the algorithmic divide in China and China-related academic and policy research in other areas. These connections are drawn out to emphasize that the algorithmic divide is a multidisciplinary research topic that can be of interest to academic and policy researchers with different training and interests.

Economic and Technological Inequalities in China

Researchers studying the algorithmic divide in China can build on, illuminate, and create synergy with China-related scholarship, policy analyses, and other research in at least three areas. The first concerns economic and technological inequalities in the country, which capture similar but broader concerns. For instance, commentators have noted the wide disparities in scientific and technological developments across China. In the intellectual property context—my primary area of research—I have observed a significant patent gap between the more economically and technologically developed provinces in the country and the less developed ones:

> In 2021, Guangdong, Jiangsu, and Zhejiang—the provinces with the three largest volumes of invention patent applications—had a total of 242,551, 188,241, and 129,821, respectively .... Meanwhile, Yunnan, Shanxi, and Guizhou (the eighteenth to twentieth provinces) had a total of only 10,293, 10,059, and 9,869, respectively. In the same year, the total number of invention patent grants for Guangdong, Jiangsu, and Zhejiang were 102,850, 68,813, and 56,796, respectively. By contrast, the total number for Yunnan, Shanxi, and Guizhou were 3,643, 3,915, and 2,824, respectively. For both applications and grants, the figures for the more developed provinces were more than twelve times the corresponding numbers for their less developed counterparts. Had we included in the second group those provinces and autonomous regions with fewer than 5,000 patent applications and 1,200 patent grants, such as Hainan, Xinjiang, Ningxia, Qinghai, and Tibet, these two groups would have even starker statistical contrasts. (Yu 2024, 90–91)

If the volumes of patent applications and grants—and, for that matter, applications and grants in other areas of intellectual property law—are indicative of the provinces' overall technological developments, the developments in those provinces with fewer than 10,000 patent applications and 3,000 grants per year clearly lag behind those provinces with more than ten times those volumes. These differentials hint at major challenges confronting policymakers. After all, it is difficult to design a nationwide innovation policy to accommodate wide subnational disparities in scientific and technological developments (Yu 2017, 2091–100).

These disparities—along with the gap in access to the Internet, ICT, digital content, and now algorithmically enhanced or AI-driven technological products and services—resonate with the growing interest among economists and policy experts in "inequality within countries." Underscoring the importance of both national and global inequality, Goal 10 of the U.N. Sustainable Development Goals calls for the reduction of "inequality [both] within and among countries," not just the latter. In the past decade, economists—most notably François Bourguignon

(2017), Branko Milanovic (2018), and Thomas Piketty (2014, 2015, 2020)—have called for greater scholarly and policy attention to the ever-increasing inequalities within countries. Based on their findings, it will not be far-fetched to predict that in the near future, inequality within countries will become a bigger issue than inequality among countries, which will continue to narrow as emerging countries move up the economic and technological ladder. As Professor Milanovic (2018, 5) declares:

> With the increases of mean incomes in Asian countries, the gaps between countries have actually been narrowing. If this trend of economic convergence continues, not only will it lead to shrinking global inequality but it will, indirectly, also give relatively greater salience to inequalities within nations.

While the level of economic and technological inequality has differed significantly from province to province, this picture does not fully capture the complexities in the spatial distribution of economic and technological activities in China. In addition to more than 20 provinces, the country also has four municipalities under the central government's direct administration—namely, Beijing, Chongqing, Shanghai, and Tianjin. Also present are many innovation hotspots in different parts of the country. In its top 100 ranking of science and technology clusters, the *Global Innovation Index 2024* report identifies more than 20 clusters within Greater China (World Intellectual Property Organization 2024b, 306–08). Among the top 20 clusters in the world are Shenzhen–Hong Kong–Guangzhou (2nd), Beijing (3rd), Shanghai–Suzhou (5th), Nanjing (9th), Wuhan (13th), Hangzhou (14th), Xi'an (18th), and Qingdao (20th) (World Intellectual Property Organization 2024b, 306).

As if the municipalities and innovation clusters had not complicated the picture of spatial distribution enough, commentators have also shown the noteworthy technological developments in certain parts of the country due to history, natural conditions, and other factors. A case in point is Darcy Pan's (2022, 2412) study of the emerging data center industry and cloud infrastructure in Guizhou, "a landlocked Southwestern province in China known for its poverty, remoteness and marginality to the rest of the country." As she observes:

> [T]he growing data center industry in Gui'an puts to use not only its natural conditions such as the climate and energy recourses, but also particular imaginaries of Guizhou's remoteness, isolation, and marginality. Government officials, industry experts and stakeholders re-use these images to reconfigure Guizhou as a naturally suitable place for data centers for foreign and domestic investors. (Pan 2022, 2423)

In sum, there are many different variables affecting the spatial distribution of scientific and technological activities in China. The study of the algorithmic divide can both strengthen and benefit from research on economic and technological inequalities in China.

Changing Innovation and Technological Landscape in China

The second area which researchers on the algorithmic divide in China can build on, illuminate, and create synergy with relates to the country's changing innovation and technological landscape, which goes hand in hand with its ability to bridge the algorithmic divide. In the past decade, China, through both the public and private sectors, has devoted growing efforts and resources to big data analytics, machine learning, and AI.

In addition, the country has undertaken a wide array of legal, policy, and institutional reforms to foster and accelerate technological developments in these areas. Particularly notable is the rise of the Cyberspace Administration of China, which now actively regulates the technology sector. Also worth highlighting is a wide array of laws and regulations that China has enacted since the mid-2000s, including the Cybersecurity Law (网络安全法) in November 2016, related provisions in the Civil Code (国民法典) in May 2020, the Data Security Law (数据安全法) in June 2021, the Personal Information Protection Law (个人信息保护法) in August 2021, in addition to the numerous judicial interpretations, implementing regulations, and measures at the provincial and municipal levels (J. Lee 2018; Lee and Zhou 2024, 244–46).

More specifically in the AI context, China has introduced regulations and measures at both the national and subnational levels. A case in point is the Interim Measures for the Management of Generative Artificial Intelligence Services (生成式人工智能服务管理暂行办法), which China adopted in July 2023 (Yu 2025). Although these measures are vague and their impact remains unclear, they aim to give China an early-mover advantage similar to the European Union's adoption of the Artificial Intelligence Act. Another example is the preparatory work that is being undertaken for the introduction of a new Artificial Intelligence Law (人工智能法), including a draft released by the Chinese Academy of Social Sciences and developed by the scholarly community. It remains to be seen how broadly drafted this statute will be and what specific language it will contain.

As if these legislative, policy, and institutional reforms were not intriguing enough, China has a strong ambition to assume global leadership in the AI space. Thus far, AI has already featured prominently in the country's strategic plans for economic, social, scientific, and technological developments (K. Lee 2018). The Next-Generation Artificial Intelligence Development Plan (新一代人工智能发展规划) states explicitly China's aim to become the world's major AI innovation center by 2030 (State Council 2017).

In the health area, China has already actively deployed AI and machine learning for many years now. As a contribution to the *Global Innovation Index 2019* report has noted, "China is turning to AI-based technologies to provide better healthcare, especially in rural areas where doctors are relying on perceptual senses, like vision and hearing, to gather information about patient health" (Khedkar and Sahay 2019, 91). In the same report, Ma Huateng (2019, 103), the CEO and co-founder of Tencent, observes:

> Th[e] growth in national health expenditures is creating opportunities for medical AI in China. According to Tractica's forecast, China's AI medical market is developing rapidly, with the market size soaring from 9.661 billion yuan in 2016, and 13.65 billion yuan in 2017, to 20.4 billion yuan in 2018, maintaining a compound annual growth rate of more than 40%. At the same time, Chinese medical institutions and businesses are taking a proactive attitude towards AI. Nearly 80% of hospitals and medical companies are planning to, or already have, carried out medical AI applications and more than 75% of hospitals believe that such applications will become popular in the future.

On AI more generally, Lee Kai-Fu (2018), the former president of Google China, recounts in his widely cited book the country's substantial engagement in this space and its active development of AI-driven products and services. With the continuous and fast-accelerating technological race between China, the European Union, and the United States, China will only devote even more resources and policy attention to boosting its internal development of big data analytics, machine learning, and AI. According to the *Artificial Intelligence Index Report 2024*, produced by the Institute for Human-Centered AI at Stanford University (2024, 14, 19), China currently dominates the world in both AI patents and installations of industrial robots. In a recent patent landscape report, the World Intellectual Property Organization (2024a, 8) also lists China-based Tencent, Ping An, Baidu, and the Chinese Academy of Sciences as the world's organizational leaders in volume of generative AI patents between 2014 and 2023. For comparison, IBM ranks only fifth in the same category.

Chinese Model of Technology Regulation

The final area which researchers on the algorithmic divide in China can build on, illuminate, and create synergy with pertains to the distinctive Chinese model of technology regulation—whether through a stand-alone or comparative analysis. Understanding this model can help promote efforts to bridge the global digital and algorithmic divides, enhance international cooperation between China and other parts of the world, and ensure greater success of technology companies conducting business in China.

In *Digital Empires*, Anu Bradford (2023) discusses the emerging battles between China, the European Union, and the United States over their influence on global technology regulation. In *High Wire*, Angela Zhang (2024) also provides a detailed account of the country's technological regulatory landscape. Given the lack of international consensus on AI and algorithmic regulation—and, for that matter, the regulation of other cutting-edge media technologies—the world's major technological powers, including China, are likely to continue to jockey for power and influence in the AI space in the next few decades.

Up until the last decade, debates on technology regulation often proceeded in a binary fashion—such as when the debate concerned the divergent privacy approaches taken by the European Union and the United States (Iowa Law Review 1995). With the ever-expanding and fast-evolving role that China is now playing in global technology regulation, these debates will likely become multipolar and pluralistic.

At the international level, one has already seen China's active engagement with norm-setting efforts in the digital trade area at the World Trade Organization and in other multilateral or plurilateral fora. Hot-button issues include cross-border transfer of information, location of computing facilities, protection for the source code of computer software, and cooperation on cybersecurity matters (Yu 2022, 757). Within China, commentators have also noted the active developments in the context of the "Digital Silk Road" (Creemers 2021; Erie and Streinz 2021)—or, more broadly, the digital or technological dimension of China's Belt and Road Initiative (Lee 2016; Yu 2019a, 2019b; Zhang and Khanal 2024).

Approaches taken by China have always garnered attention from Western observers, due partly to the country's drastically different political system and information control environment and partly to its widely criticized state-driven innovation practices. Since China's reopening to the outside world in the late 1970s, policymakers, commentators, and the mass media have devoted little attention to the country's technological developments. Many are now paying greater attention, partly out of concern that such developments will have major negative ramifications for economic and technological developments in their home countries as well as in third countries. Immediately coming to mind are the heated debates in the United States and other countries about the security threats posed by Huawei Technologies and ByteDance (the developer of video-sharing app TikTok) and the increasing economic and technological dominance of China-based Alibaba and Tencent.

In emerging areas such as ADM and automated governance—and, more broadly, in the development of an algorithmic society—comparative research can be highly beneficial. To accommodate local needs, interests, conditions, and priorities, countries understandably adopt different approaches to address opportunities and challenges unleashed by advances in big data analytics, machine learning, and

AI. Because these advances can affect issues such as national security and global competitiveness, countries remain highly reluctant to forgo their first-mover or comparative advantage by accepting models developed abroad, which are often designed with different local conditions in mind.

In sum, highlighting the strengths and weaknesses of different approaches to technology regulation taken by China and other parts of the world will deepen our appreciation of the difficult choices policymakers have to make in response to changes precipitated by advances in big data analytics, machine learning, and AI. Such comparative analyses will also allow us to better prepare for the unintended consequences and spillover effects sparked by these policy choices.

Illustrations from Law and Policy Responses

The previous section has identified three areas which researchers on the algorithmic divide in China can build on, illuminate, and create synergy with. To highlight the comparative insights provided by studying this divide, this chapter examines three sets of policy responses that commentators have proposed to bridge the algorithmic divide. For analytical convenience and effectiveness, these policy responses are drawn from legal and policy literature, with which I am most familiar. Nevertheless, it is important to keep in mind that these responses are provided as mere illustrations. Because the algorithmic divide, like the digital divide, is a "multidimensional phenomenon" (Norris 2001, 4; Van Dijk 2005, 3), any effort to bridge the algorithmic divide is likely to require a holistic effort that involves many different policy responses.

Enhance Algorithmic Literacy

The first set of policy responses seeks to enhance algorithmic literacy. From a lack of awareness of algorithmically related problems to an inability to adapt to machine learning and AI (Yu 2020c, 338–39, 342–43), increasing such literacy is crucial to ensure a large majority of the population reap the benefits of these new technologies (IEEE 2017, 142; Paul, Jolley, and Anthony 2018, 74; Rainie and Anderson 2017, 74–76; UNESCO 2019, 6–7, 29).

To be algorithmically literate, individuals need to know not only the impact of machine learning and AI on daily lives but also what it means to live in an algorithmic society (Denning and Tedre 2019; International Society for Technology in Education and Computer Science Teachers Association n.d.; Warschauer 2003, 113). Greater algorithmic literacy will help these individuals realize the full potential of machine learning and AI. It will also assist them in avoiding undesirable technological products and services that fail to protect privacy or other fundamental rights. In addition, a greater understanding of algorithmic operations will allow individuals to develop human-generated responses to ensure more successful engagement with algorithmically enhanced or AI-driven technological products and services.

While most individuals are unlikely to be able to fully understand the algorithmic operations involved (Chander 2017, 1040; Domingos 2012, 5; Eubanks 2017, 184–85; Kroll et al. 2017, 638; Rainie and Anderson 2017, 19)—or, in some cases, no individual will ever be able to develop such a full understanding—research has shown that individuals are capable of developing responses that will "trick" algorithms into providing more desirable results. For instance, Facebook users have provided different information to improve algorithmic outcomes (Bambauer and Zarsky 2018, 12–14; Garling 2014; Susarla 2019). Research has also shown that users change their behaviors to avoid undesirable outcomes (Burk 2019, 295; Yu 2020a, 200–02; 2020b, 334–5). Having strong algorithmic literacy is therefore crucial to adapting to this new technological environment, not to mention an individual's general need for such literacy to avoid job displacement (Brynjolfsson and McAfee 2014; Estlund 2018; Executive Office of the President 2016, 35; Hamann 2018; Rainie and Anderson 2017, 70–73; Yu 2020c, 353–54).

In China specifically, it may be useful to examine the variations in algorithmic literacy in different parts of the country. For example, how significant are these variations from province to province, across gender and age groups, and based on educational backgrounds, income levels, and technological competence? It will also be instructive to compare algorithmic literacy in China with those in other parts of the world or at the global level. In addition, it can be helpful to explore whether the greater recent deployment of technological tools, such as during the COVID-19 pandemic, has altered these comparative findings. One could further interrogate whether the level of algorithmic literacy has risen significantly in the past decade, due to the country's increased policy attention on and state-driven investments in big data analytics, machine learning, and AI.

Increase Accountability and Responsiveness

The second set of policy responses aims to increase accountability and responsiveness. Thus far, many commentators have underscored the need for greater transparency and accountability in the design and use of algorithms (Kroll et al. 2017; Pasquale 2019b), including the disclosure of technological choices made by algorithm designers (Bloch-Wehba 2020, 1295–306; Citron 2008, 371–81; Katyal 2019b, 1250–79). As a group of legal and computer science researchers declare, "in order for a computer system to function in an accountable way—either while operating an important civic process or merely engaging in routine commerce—accountability must be part of the system's design from the start" (Kroll et al. 2017, 640). Some experts and professional associations have gone even further to call on operators of ADM systems to be subject to periodic assessments—either conducted internally or through external audits (Center for Democracy and Technology 2018, 11; Desai and Kroll 2017, 36–42; Diakopoulos et al. n.d.; Kim 2017a, 190–91; Yu 2020c, 380–82). Their calls for periodic assessments underscore the need for evaluations at different stages of the algorithm design and development process (Diakopoulos et al. n.d.).

Notwithstanding these proposals and other similar policy responses, building accountability and transparency into an environment involving AI and machine learning is not easy. To begin with, algorithmic transparency requires the disclosure of not only the algorithms involved (and the accompanying source code), but also training data and algorithmic outcomes (Chander 2017, 1024–25; Kroll et al. 2017, 641; O'Neil 2016, 229). It is particularly important to disclose these outcomes because many of them will re-enter ADM systems as training or feedback data. The continuous provision of these data will therefore create a self-reinforcing feedback loop that amplifies the "garbage in, garbage out" problem, turning inaccurate, biased, or otherwise inappropriate inputs into faulty outputs (Center for Democracy and Technology 2018, 8; Grafanaki 2017, 827; Katyal 2019a, 69; Yu and Spina Alì 2019, 4). As time passes, the biases, bugs, and other problems generated through these feedback loops will become much worse than those found in the original algorithm designs or initial training data.

Worse still, it remains unclear if the full disclosure of all the information involved in the design and operation of the algorithms will allow users or their advocates to identify the problem. For instance, such disclosure may result in an unmanageable deluge of information, making it very difficult, if not impossible, for the public to understand how data are used and how these algorithms generate outcomes (Cohen 2019, 180; Perel and Elkin-Koren 2017, 194–96). Many commentators have also lamented how the public often finds source code and training data incomprehensible (Chander 2017, 1040; Kroll et al. 2017, 638; Noto La Diega 2018, 23; Rainie and Anderson 2017, 19). Even for those having the requisite skills to handle computer code and technical data, analyzing all the disclosed information will require considerable time, effort, resources, and energy (Perel and Elkin-Koren 2017, 195–96; Yu 2020c, 375). Such analysis will therefore be cost-prohibitive and difficult to conduct, except for individual projects.

In China, it has always been challenging to discuss issues relating to accountability and transparency, due in large part to the country's political system, administrative environment, and information control practices. Unlike the calls for transparency in the ADM context in the United States, similar calls in China are unlikely to bring to mind Justice Louis Brandeis's (1913, 10) century-old adage that "[s]unlight is said to be the best of disinfectants." Nevertheless, many approaches are still available in the country to promote accountability and transparency. For example, in addition to public disclosure, commentators have called for the utilization of external auditors (Yu 2020c, 380–81), ombudspersons (McGregor et al. 2019, 332), or oversight bodies (IEEE 2017, 70; Pasquale 2011, 247). Some of these recommendations can be implemented in China.

Moreover, one should avoid equating transparency with accountability (Perel and Elkin-Koren 2017, 184). As the political processes in the European Union, the United States, and other parts of the world have repeatedly demonstrated, one could have a highly transparent process that involves checks and balances, different rounds of open consultations, and a large number of publicly available documents, yet the outcomes remain heavily captured by industries and are of limited public accountability (Horten 2013; Lindsey and Teles 2017, 64–89). It is therefore important to keep in mind that a key objective of the efforts to promote accountability and transparency in the ADM context is to ensure that operators can be held responsible for problems arising in their systems. Through these and other efforts, such operators can be induced, either proactively or preemptively, to introduce safeguards and remedies (Association for Computing Machinery 2017, 1–2).

One preemptive measure that many commentators have supported is the development of explainable algorithms (Association for Computing Machinery 2017, 2; Diakopoulos et al. n.d.; IEEE 2017, 68; Kaminski 2019; Selbst and Barocas 2018; Selbst and Powles 2017; Yu 2020c, 377–78). As Pauline Kim (2017, 922–23) elaborates:

> When a model is interpretable, debate may ensue over whether its use is justified, but it is at least possible to have a conversation about whether relying on the behaviors or attributes that drive the outcomes is normatively acceptable. When a model is not interpretable, however, it is not even possible to have the conversation.

The development of explainable algorithms and other similar measures is certainly something that can be implemented in China. Such development is also important to the country, as it will increase trust in algorithms and ADM systems, thereby supporting an expansion of automated governance.

Manage the Human–Machine Interface

The last set of policy responses targets the interface between humans and machines—and, in the context of this volume, the interface between ADM systems and their human operators. From both a policy and research standpoint, it will be worthwhile to explore the readiness of these systems for human intervention (Huq 2020; Jones 2017; Roberts 2019; Yu 2021, 285–90). Such intervention—or what some commentators have referred to as "human in the loop" (Buckley et al. 2021; Jones 2017)—is important considering that humans are known to have made better decisions than machines in many situations, especially unprecedented ones (Agrawal et al. 2018, 59; Pasquale 2019a, 53). As Anthony Casey and Anthony Niblett (2019, 354) point out:

> Algorithmic decision-making does not mean that humans are shut out of the process. Even after the objective has been set, there is much human work to be done. Indeed, humans are involved in all stages of setting up, training, coding, and assessing the merits of the algorithm. If the objectives of the algorithm and the objective of the law are perfectly aligned at the ex ante stage, one must ask: Under what circumstances should a human ignore the algorithm's suggestions and intervene *after* the algorithm has made the decision?

Moreover, increasing the ADM systems' readiness for human intervention will enable decisions made by human operators to be fed back into the algorithms as training and feedback data. Such intervention will therefore make the systems even better and more robust.

An instructive example is the debacle confronting Uber when a gunman took 17 hostages at the Lindt Chocolate Café in Sydney, Australia, in December 2014. Because many people were trying to flee the Central Business District simultaneously, the sudden increase in demand for rideshares caused the platform to "impose[] surge pricing in the city, charging passengers a minimum of [AUD] 100 for a ride, four times the normal fare" (Lapowsky 2014). Unfortunately, the pricing algorithm was unable to connect the dots the same way a human operator could, especially after the tragic news about the hostages began to pour in from the Internet and other traditional media.

Even worse for Uber, charging higher prices in such an emergency situation created bad public relations—not that different from our reactions to opportunistic price surges among retailers in the first few months of the COVID-19 pandemic. Following the unfortunate episode in Sydney, Uber quickly issued an apology and offered refunds to affected customers. Having learned a lesson, it also "put in place the ability to override automatic surge pricing in some circumstances" (McAfee and Brynjolfsson 2017, 55). By the time a series of terrorist attacks broke out in Paris a year later, Uber was able to "cancel[] surge pricing in the city [within half an hour of the first attack] and alerted all of its users to the emergency" (McAfee and Brynjolfsson 2017, 55). This drastically different outcome shows the importance and wisdom of increasing an ADM system's readiness for human intervention.

While human operators often intervene based on internal data, they can also rely on external information. In an earlier article, I advocated the development of a notice-and-correct mechanism to rectify problems generated by ADM systems and algorithmically enhanced or AI-driven technological products and services (Yu 2020c, 379–80). Inspired by the notice-and-takedown arrangements in copyright law, my proposed mechanism underscores the need for operators of these systems to take expedited actions after they have been notified of problems generated by the algorithms used in the systems (Association for Computing Machinery 2017, 2; Brownsword 2019, 297; Chander 2017, 1025; Crawford and Schultz 2014; Diakopoulos et al. n.d.; Yu 2020c, 379–80). As I explained, "as technology becomes increasingly complicated and inscrutable, ensuring quick correction of the problem will likely be more constructive than punishing those who have allowed the problems to surface in the first place, often unintentionally" (Yu 2020c, 380).

Notwithstanding the need for and benefits of human intervention, deciding whether and when to intervene can be difficult, especially in an environment involving AI and machine learning. While developers of ADM systems can set up monitoring procedures to ensure that the algorithmically generated outcomes from these systems match human intuition, such procedures may undermine a key advantage of machine learning and AI. Because humans and machines "think" differently (Millar and Kerr 2016, 117–24), these systems can generate seemingly counterintuitive decisions that are superior to human decisions (Center for Democracy and Technology 2018, 2; Guo and Li 2018, 175; Kurzweil 2005, 261; Topol 2019, 117–18; Yu 2020a, 215–16). Even more challenging, human operators, due to cognitive barriers, may not always be able to fully appreciate the merits of those counterintuitive decisions. As Professors Casey and Niblett (2019, 354) observe:

> Algorithms will often identify counterintuitive connections that may appear erroneous to humans even when accurate. Humans should be careful in those cases not to undo the very value that was added by the algorithm's ability to recognize these connections. This is especially true when the benefit of the algorithm was that it reduced human bias and behavioral errors.

Thus, as important as it is for human operators of ADM systems to intervene, they should be careful not to quickly reject counterintuitive algorithmically generated outcomes (Rainie and Anderson 2017, 40; Selbst and Barocas 2018; Surden and Williams 2016, 158). What looks counterintuitive at first glance may make more sense with hindsight.

In China, it is easily understandable that some would assume that the choice over whether human operators of ADM systems should intervene or defer may differ significantly from the choices made in other jurisdictions, such as the European Union or the United States. Yet, based on what we have seen so far, choices concerning how to manage the human–machine interface are so novel and technologically challenging that virtually all countries are now scrambling to develop guidelines to help make these difficult choices. How China handles the human–machine interface—or, more specifically, the interface between ADM systems and their human operators—will inform the experiences of other countries, and vice versa.

Conclusion

Since the mid-1990s, the digital divide has left behind many individuals, taking away their political, social, economic, cultural, educational, and career opportunities. With the arrival of big data analytics, machine learning, and AI, the algorithmic divide threatens to leave behind even more people, not only those who are already struggling on the unfortunate side of the digital divide.

Focusing on a topic that has been underexplored in ADM literature, this chapter explains why the algorithmic divide in China can provide an important multidisciplinary case study at both the comparative and global levels. Even better, by studying this divide in China, we can gain valuable insights into the country's economic and technological inequalities, changing innovation and technological landscape, and distinctive model of technology regulation.

It remains to be seen how quickly we can bridge the algorithmic divide in China or in other parts of the world. Even if we manage to narrow it significantly, it is possible that we may never be able to eliminate this inequitable gap. Nevertheless, if this chapter can inspire policy and academic experts to undertake greater research on the algorithmic divide in China, it will have accomplished its primary objective. With the growing use of ADM systems in today's algorithmic society, understanding this divide and developing solutions to address it is not only important and beneficial but also urgent.

Acknowledgements

This chapter draws on, and adapts, research the author conducted for articles published in the *Florida Law Review* and the *Northeastern University Law Review*. The author is grateful to the editors of this volume, two anonymous reviewers, and Elisa Oreglia for their helpful comments and suggestions.

References

Agrawal, Ajay, Joshua Gans, and Avi Goldfarb. 2018. *Prediction Machines: The Simple Economics of Artificial Intelligence*. Boston: Harvard Business Review Press.

Association for Computing Machinery, U.S. Public Policy Council. 2017. "Statement on Algorithmic Transparency and Accountability." https://www.acm.org/binaries/content/assets/public-policy/2017_usacm_statement_algorithms.pdf.

Balkin, Jack M. 2017. "2016 Sidley Austin Distinguished Lecture on Big Data Law and Policy: The Three Laws of Robotics in the Age of Big Data." *Ohio State Law Journal* 78: 1217–41.

Bambauer, Jane, and Tal Zarsky. 2018. "The Algorithm Game." *Notre Dame Law Review* 94: 1–47.

Barocas, Solon, and Andrew D. Selbst. 2016. "Big Data's Disparate Impact." *California Law Review*. 104: 671–732.

Bentley, Sarah V., Claire K. Naughtin, Melanie J. McGrath, Jessica L. Irons, and Patrick S. Cooper. 2024. "The Digital Divide in Action: How Experiences of Digital Technology Shape Future Relationships with Artificial Intelligence." *AI Ethics* 4: 901–15.

Bloch-Wehba, Hannah. 2020. "Access to Algorithms." *Fordham Law Review* 88: 1265–314.

Bornstein, Stephanie. 2018. "Antidiscriminatory Algorithms." *Alabama Law Review* 70: 519–72.

Bourguignon, François. 2017. *The Globalization of Inequality*, translated by Thomas Scott-Railton. Princeton: Princeton University Press.

Bradford, Anu. 2023. *Digital Empires: The Global Battle to Regulate Technology*. Oxford: Oxford University Press.

Brandeis, Louis D. 1913. "What Publicity Can Do." *Harper's Weekly*, December 20, 10–13.

Brownsword, Roger. 2019. *Law, Technology and Society: Re-Imagining the Regulatory Environment*. Abingdon: Routledge.

Brynjolfsson, Erik, and Andrew McAfee. 2014. *The Second Machine Age: Work, Progress, and Prosperity in a Time of Brilliant Technologies*. New York: W. W. Norton & Co.

Buckley, Ross P., Dirk A. Zetzsche, Douglas W. Arner, and Brian W. Tang. 2021. "Regulating Artificial Intelligence in Finance: Putting the Human in the Loop." *Sydney Law Review* 43: 43–81.

Burk, Dan L. 2019. "Algorithmic Fair Use." *University of Chicago Law Review* 86: 283–307.

Casey, Anthony J., and Anthony Niblett. 2019. "A Framework for the New Personalization of Law." *University of Chicago Law Review* 86: 333–58.

Center for Democracy and Technology. 2018. "Digital Decisions." https://cdt.org/files/2018/09/Digital-Decisions-Library-Printer-Friendly-as-of-20180927.pdf.

Chander, Anupam. 2017. "The Racist Algorithm?" *Michigan Law Review* 115: 1023–45.

China Internet Network Information Center (CNNIC). 2024. *The 54th Statistical Report on China's Internet Development*. Beijing: China Internet Network Information Center.

Citron, Danielle Keats. 2008. "Open Code Governance." *University of Chicago Legal Forum* 2008: 355–87.

Cohen, Julie E. 2019. *Between Truth and Power: The Legal Constructions of Informational Capitalism*. New York: Oxford University Press.

Compaine Benjamin M., ed. 2001. *The Digital Divide: Facing a Crisis or Creating a Myth?* Cambridge, Mass.: MIT Press.

Crawford, Kate, and Jason Schultz. 2014. "Big Data and Due Process: Toward a Framework to Redress Predictive Privacy Harms." *Boston College Law Review* 55: 93–128.

Creemers, Rogier, ed. 2021. *The Digital Silk Road: Perspectives from Affected Countries*. Leiden: Leiden Asia Centre.

De-Arteaga, Maria, William Herlands, Daniel B. Neill, and Artur Dubrawski. 2018. “Machine Learning for the Developing World.” *ACM Transactions on Management Information Systems* 2018, no. 9: 1–14.

Denning, Peter J., and Matti Tedre. 2019. *Computational Thinking*. Cambridge, Mass.: MIT Press.

Desai, Deven R., and Joshua A. Kroll. 2017. “Trust but Verify: A Guide to Algorithms and the Law.” *Harvard Journal of Law and Technology* 31: 1–64.

Diakopoulos, Nicholas, Sorelle Friedler, Marcelo Arenas, Solon Barocas, Michael Hay, Bill Howe, H. V. Jagadish, Kris Unsworth, Arnaud Sahuguet, Suresh Venkatasubramanian, Christo Wilson, Cong Yu, and Bendert Zevenbergen. n.d. “Principles for Accountable Algorithms and a Social Impact Statement for Algorithms.” Accessed June 10, 2024. https://www.fatml.org/resources/principles-for-accountable-algorithms.

Domingos, Pedro. 2015. *The Master Algorithm: How the Quest for the Ultimate Learning Machine Will Remake Our World*. New York: Basic Books.

Erie, Matthew S., and Thomas Streinz. 2021. “The Beijing Effect: China’s Digital Silk Road as Transnational Data Governance.” *New York University Journal of International Law and Politics* 54: 1–92.

Estlund, Cynthia. 2018. “What Should We Do after Work? Automation and Employment Law.” *Yale Law Journal* 128: 254–326.

Executive Office of the President (U.S.). 2016. *Artificial Intelligence, Automation, and the Economy*. Washington, D.C.: Executive Office of the President.

Eubanks, Virginia. 2017. *Automating Inequality: How High-Tech Tools Profile, Police, and Punish the Poor*. New York: St. Martin’s Press.

Frischmann, Brett, and Evan Selinger. 2018. *Re-Engineering Humanity*. Cambridge: Cambridge University Press.

Garling, Caleb. 2014. “Tricking Facebook’s Algorithm.” *Atlantic*, August 8. https://www.theatlantic.com/technology/archive/2014/08/tricking-facebooks-algorithm/375801/.

Goggin, Gerard, and Karen Soldatić. 2022. “Automated Decision-Making, Digital Inclusion and Intersectional Disabilities.” *New Media and Society* 24: 384–400.

Grafanaki, Sofia. 2017. “Autonomy Challenges in the Age of Big Data.” *Fordham Intellectual Property, Media and Entertainment Law Journal* 27: 803–68.

Guo, Jonathan, and Li Bin. 2018. “The Application of Medical Artificial Intelligence Technology in Rural Areas of Developing Countries.” *Health Equity* 2: 174–81.

Hamann, Ralph. 2018. “Developing Countries Need to Wake Up to the Risks of New Technologies.” *The Conversation*, January 4. https://theconversation.com/developing-countries-need-to-wake-up-to-the-risks-of-new-technologies-87213.

Hargittai, Eszter, ed. 2021. *Handbook of Digital Inequality*. Cheltenham: Edward Elgar Publishing.

Horten, Monica. 2013. *A Copyright Masquerade: How Corporate Lobbying Threatens Online Freedoms*. London: Zed Books.

Huq, Aziz Z. 2019. “Racial Equity in Algorithmic Criminal Justice.” *Duke Law Journal* 68: 1043–134.

Huq, Aziz Z. 2020. “A Right to a Human Decision.” *Virginia Law Review* 106: 611–88.

Institute of Electrical and Electronics Engineers (IEEE). 2017. *Ethically Aligned Design: A Vision for Prioritizing Human Well-Being with Autonomous and Intelligent Systems*. New York: Institute of Electrical and Electronics Engineers.

International Society for Technology in Education and Computer Science Teachers Association. n.d. “Operational Definition of Computational Thinking for K-12 Education.” https://cdn.iste.org/www-root/Computational_Thinking_Operational_Definition_ISTE.pdf.

Iowa Law Review. 1995. “Data Protection Law and the European Union’s Directive: The Challenge for the United States.” *Iowa Law Review* 80: 431–695.

Jarrett, Cosette. 2017. "AI Could Be Driving a New Digital Divide." *Venturebeat*, October 5. https://venturebeat.com/2017/10/05/ai-could-be-driving-a-new-digital-divide/.

Jones, Meg Leta. 2017. "The Right to a Human in the Loop: Political Constructions of Computer Automation and Personhood." *Social Studies of Science* 47: 216–39.

Kaminski, Margot E. 2019. "The Right to Explanation, Explained." *Berkeley Technology Law Journal* 34: 189–218.

Katyal, Sonia K. 2019a. "Private Accountability in the Age of Artificial Intelligence." *UCLA Law Review* 66: 54–141.

Katyal, Sonia K. 2019b. "The Paradox of Source Code Secrecy." *Cornell Law Review* 104: 1183–279.

Khedkar, Pratap, and Dharmendra Sahay. 2019. "Trends in Healthcare and Medical Innovation." In *The Global Innovation Index 2019: Creating Healthy Lives—The Future of Medical Innovation*, edited by Soumitra Dutta, Bruno Lanvin, and Sacha Wunsch-Vincent, 87–93. Ithaca, Fontainebleau, and Geneva: Cornell University, INSEAD, and the World Intellectual Property Organization.

Kim, Pauline T. 2017a. "Auditing Algorithms for Discrimination." *University of Pennsylvania Law Review Online* 166: 189–203.

Kim, Pauline T. 2017b. "Data-Driven Discrimination at Work." *William and Mary Law Review* 58: 857–936.

Kroll, Joshua A., Joanna Huey, Solon Barocas, Edward W. Felten, Joel R. Reidenberg, David G. Robinson, and Harlan Yu. 2017. "Accountable Algorithms." *University of Pennsylvania Law Review* 165: 633–705.

Kurzweil, Ray. 2005. *The Singularity Is Near: When Humans Transcend Biology*. New York: Penguin Books.

Lapowsky, Issie. 2014. "What Uber's Sydney Surge Pricing Debacle Says about Its Public Image." *Wired*, December 15. https://www.wired.com/2014/12/uber-surge-sydney/.

Lee, Jyh-An. 2016. "The New Silk Road to Global IP Landscape." In *Legal Dimensions of China's Belt and Road Initiative*, edited by Lutz-Christian Wolff and Xi Chao, 417–30. Hong Kong: Wolters Kluwer Hong Kong Limited.

Lee, Jyh-An. 2018. "Hacking into China's Cybersecurity Law." *Wake Forest Law Review* 53: 57–104.

Lee, Jyh-An, and Zhou Peng. 2024. "FRT Regulation in China." In *The Cambridge Handbook of Facial Recognition in the Modern State*, edited by Rita Matulionyte and Monika Zalnieriute, 242–52. Cambridge: Cambridge University Press.

Lee, Kai-Fu. 2018. *AI Superpowers: China, Silicon Valley, and the New World Order*. Boston: Houghton Mifflin Harcourt.

Lindsey, Brink, and Steven M. Teles. 2017. *The Captured Economy: How the Powerful Enrich Themselves, Slow Down Growth, and Increase Inequality*. New York: Oxford University Press.

Louie, Scott. 1999. "The New Invisible Man." *Yo! Youth Outlook*, November 1.

Ma, Huateng. 2019. "Application of Artificial Intelligence and Big Data in China's Healthcare Services." In *The Global Innovation Index 2019: Creating Healthy Lives—The Future of Medical Innovation*, edited by Soumitra Dutta, Bruno Lanvin, and Sacha Wunsch-Vincent, 103–09. Ithaca, Fontainebleau, and Geneva: Cornell University, INSEAD, and the World Intellectual Property Organization.

Mack, Raneta Lawson. 2001. *The Digital Divide: Standing at the Intersection of Race & Technology*. Durham: Carolina Academic Press.

McAfee, Andrew, and Erik Brynjolfsson. 2017. *Machine, Platform, Crowd: Harnessing Our Digital Future*. New York: W. W. Norton & Co.

McGregor, Lorna, Daragh Murray, and Vivian Ng. 2019. "International Human Rights Law as a Framework for Algorithmic Accountability." *International and Comparative Law Quarterly* 68: 309–43.

Michaels, Andrew C. 2020. "Artificial Intelligence, Legal Change, and Separation of Powers." *University of Cincinnati Law Review* 88: 1083–103.

Milanovic, Branko. 2018. *Global Inequality: A New Approach for the Age of Globalization*. Cambridge, Mass.: Belknap Press of Harvard University Press.

Millar, Jason, and Ian Kerr. 2016. "Delegation, Relinquishment and Responsibility: The Prospect of Expert Robots." In *Robot Law*, edited by Ryan Calo, A. Michael Froomkin, and Ian Kerr, 102–28. Cheltenham: Edward Elgar Publishing.

National Telecommunications and Information Administration. 1995. *Falling through the Net: A Survey of the "Have Nots" in Rural and Urban America*. Washington, D.C.: U.S. Department of Commerce.

National Telecommunications and Information Administration. 1998. *Falling through the Net II: New Data on the Digital Divide*. Washington, D.C.: U.S. Department of Commerce.

National Telecommunications and Information Administration. 1999. *Falling through the Net: Defining the Digital Divide*. Washington, D.C.: U.S. Department of Commerce.

National Telecommunications and Information Administration. 2000. *Falling through the Net: Toward Digital Inclusion*. Washington, D.C.: U.S. Department of Commerce.

Norris, Pippa. 2001. *Digital Divide: Civic Engagement, Information Poverty, and the Internet Worldwide*. Cambridge: Cambridge University Press.

Noto La Diega, Guido. 2018. "Against the Dehumanisation of Decision-Making—Algorithmic Decisions at the Crossroads of Intellectual Property, Data Protection, and Freedom of Information." *Journal of Intellectual Property, Information Technology and Electronic Commerce Law* 9: 3–34.

O'Neil, Cathy. 2016. *Weapons of Math Destruction: How Big Data Increases Inequality and Threatens Democracy*. New York: Crown.

Packin, Nizan, and Yafit Lev-Aretz. 2018. "Learning Algorithms and Discrimination." In *Research Handbook on the Law of Artificial Intelligence*, edited by Woodrow Barfield and Ugo Pagallo, 88–113. Cheltenham: Edward Elgar Publishing.

Pan, Darcy. 2022. "Storing Data on the Margins: Making State and Infrastructure in Southwest China." *Information, Communication and Society* 25: 2412–26.

Pasquale, Frank. 2011. "Restoring Transparency to Automated Authority." *Journal on Telecommunications and High Technology Law* 9: 235–54.

Pasquale, Frank. 2015. *The Black Box Society: The Secret Algorithms That Control Money and Information*. Cambridge, MA, USA: Harvard University Press.

Pasquale, Frank. 2019a. "A Rule of Persons, Not Machines: The Limits of Legal Automation." *George Washington Law Review* 87: 1–55.

Pasquale, Frank. 2019b. "The Second Wave of Algorithmic Accountability." *Law and Political Economy*, November 25. https://lpeblog.org/2019/11/25/the-second-wave-of-algorithmic-accountability/.

Paul, Amy, Craig Jolley, and Aubra Anthony. 2018. *Reflecting the Past, Shaping the Future: Making AI Work for International Development*. Washington, D.C.: U.S. Agency for International Development.

Peacock, Anne. 2019. *Human Rights and the Digital Divide*. Abingdon: Routledge.

Perel, Maayan, and Niva Elkin-Koren. 2017. "Black Box Tinkering: Beyond Disclosure in Algorithmic Enforcement." *Florida Law Review* 69: 181–221.

Pick, James B., and Avijit Sarkar, eds. 2015. *The Global Digital Divides: Explaining Change*. Heidelberg: Springer.

Piketty, Thomas. 2014. *Capital in the Twenty-First Century*, translated by Arthur Goldhammer. Cambridge, Mass.: Harvard University Press.

Piketty, Thomas. 2015. *The Economics of Inequality*, translated by Arthur Goldhammer. Cambridge, Mass.: The Belknap Press of Harvard University Press.

Piketty, Thomas. 2020. *Capital and Ideology*, translated by Arthur Goldhammer. Cambridge, Mass.: The Belknap Press of Harvard University Press.

Ragnedda, Massimo. 2017. *The Third Digital Divide: A Weberian Approach to Digital Inequalities*. Abingdon: Routledge.

Ragnedda, Massimo, and Glenn W. Muschert, eds. 2013. *The Digital Divide: The Internet and Social Inequality in International Perspective*. Abingdon: Routledge.

Ragnedda, Massimo, and Glenn W. Muschert, eds. 2018. *Theorizing Digital Divides*. Abingdon: Routledge.

Rainie, Lee, and Janna Anderson. 2017. *Code-Dependent: Pros and Cons of the Algorithm Age*. Washington, D.C.: Pew Research Center.

Roberts, Sarah T. 2019. *Behind the Screen: Content Moderation in the Shadows of Social Media*. New Haven: Yale University Press.

Selbst, Andrew D. 2017. "Disparate Impact in Big Data Policing." *Georgia Law Review* 52: 109–95.

Selbst, Andrew D., and Solon Barocas. 2018. "The Intuitive Appeal of Explainable Machines." *Fordham Law Review* 87: 1085–139.

Selbst, Andrew D., and Julia Powles. 2017. "Meaningful Information and the Right to Explanation." *International Data Privacy Law* 7: 233–42.

Shao, Qinglong, and Genia Kostka. 2023. "The COVID-19 Pandemic and Deepening Digital Inequalities in China." *Telecommunications Policy* 47: 102644.

Song, Zhouying, Wang Chen, and Luke Bergmann. 2020. "China's Prefectural Digital Divide: Spatial Analysis and Multivariate Determinants of ICT Diffusion." *International Journal of Information Management* 52: 102072.

Stanford University, Institute for Human-Centered AI. 2024. *The AI Index 2024 Annual Report*. Palo Alto: Stanford University.

State Council of the People's Republic of China. 2017. "Notice of the Next-Generation Artificial Intelligence Development Plan." Notice No. 35.

Surden, Harry, and Mary-Anne Williams. 2016. "Technological Opacity, Predictability, and Self-Driving Cars." *Cardozo Law Review* 38: 121–81.

Susarla, Anjana. 2019. "The New Digital Divide Is between People Who Opt Out of Algorithms and People Who Don't." *The Conversation*, April 17. https://theconversation.com/the-new-digital-divide-is-between-people-who-opt-out-of-algorithms-and-people-who-dont-114719.

Topol, Eric J. 2019. *Deep Medicine: How Artificial Intelligence Can Make Healthcare Human Again*. New York: Basic Books.

United Nations. 2020. "Digital Divide 'a Matter of Life and Death' amid COVID-19 Crisis, Secretary-General Warns Virtual Meeting, Stressing Universal Connectivity Key for Health, Development." June 11. https://press.un.org/en/2020/sgsm20118.doc.htm.

United Nations Educational, Scientific and Cultural Organization (UNESCO). 2019. *Artificial Intelligence in Education: Challenges and Opportunities for Sustainable Development*. Paris: United Nations Educational, Scientific and Cultural Organization.

United Nations Secretary-General's High-Level Advisory Body on Artificial Intelligence (U.N. High-Level Advisory Body). 2024. *Governing AI for Humanity: Final Report*. New York: United Nations.

United Nations Secretary-General's High-Level Panel on Digital Cooperation (U.N. High-Level Panel). 2019. *The Age of Digital Interdependence*. New York: United Nations.

Van Dijk, Jan A.G.M. 2005. *The Deepening Divide: Inequality in the Information Society*. Thousand Oaks: Sage Publications.

Van Dijk, Jan. 2020. *The Digital Divide*. Cambridge: Polity Press.

Warschauer, Mark. 2003. *Technology and Social Inclusion: Rethinking the Digital Divide*. Cambridge, Mass.: MIT Press.

World Intellectual Property Organization. 2024a. *Generative Artificial Intelligence: Patent Landscape Report*. Geneva: World Intellectual Property Organization.

World Intellectual Property Organization. 2024b. *Global Innovation Index 2024: Unlocking the Promise of Social Entrepreneurship*, edited by Soumitra Dutta, Bruno Lanvin, Lorena Rivera León, and Sacha Wunsch-Vincent. Geneva: World Intellectual Property Organization.

Yang, Guobin. 2022. *The Wuhan Lockdown*. New York: Columbia University Press.

Yu, Haiqing. 2024. "Living in the Era of Codes: A Reflection on China's Health Code System." *Biosocieties* 19: 1–18.

Yu, Peter K. 2002. "Bridging the Digital Divide: Equality in the Information Age." *Cardozo Arts and Entertainment Law Journal* 20: 1–52.

Yu, Peter K. 2012. "Intellectual Property and Asian Values." *Marquette Intellectual Property Law Review* 16: 329–99.

Yu, Peter K. 2014. "The Middle Intellectual Property Powers." In *Law and Development of Middle-Income Countries: Avoiding the Middle-Income Trap*, edited by Randall Peerenboom and Tom Ginsburg, 84–107. New York: Cambridge University Press.

Yu, Peter K. 2017. "A Spatial Critique of Intellectual Property Law and Policy." *Washington and Lee Law Review* 74: 2045–132.

Yu, Peter K. 2019a. "Building Intellectual Property Infrastructure along China's Belt and Road." *University of Pennsylvania Asian Law Review* 14: 275–325.

Yu, Peter K. 2019b. "China, 'Belt and Road' and Intellectual Property Cooperation." *Global Trade and Customs Journal* 14: 244–50.

Yu, Peter K. 2020a. "Artificial Intelligence, the Law–Machine Interface, and Fair Use Automation." *Alabama Law Review* 72: 187–238.

Yu, Peter K. 2020b. "Can Algorithms Promote Fair Use?" *FIU Law Review* 14: 329–63.

Yu, Peter K. 2020c. "The Algorithmic Divide and Equality in the Age of Artificial Intelligence." *Florida Law Review* 72: 331–89.

Yu, Peter K. 2021. "Beyond Transparency and Accountability: Three Additional Features Algorithm Designers Should Build into Intelligent Platforms." *Northeastern University Law Review* 13: 263–96.

Yu, Peter K. 2022. "Fitting Machine-Generated Data into Trade Regulatory Holes." In *Trade in Knowledge: Intellectual Property, Trade and Development in a Transformed Global Economy*, edited by Antony Taubman and Jayashree Watal, 738–67. Cambridge: Cambridge University Press.

Yu, Peter K. 2024. "Intellectual Property, Global Inequality, and Subnational Policy Variations." In *Intellectual Property, Innovation and Economic Inequality*, edited by Daniel Benoliel, Peter K. Yu, Francis Gurry, and Keun Lee, 81–105. Cambridge: Cambridge University Press.

Yu, Peter K. 2025. "Artificial Intelligence, Autonomous Creation, and the Future Path of Copyright Law." *Brigham Young University Law Review* 50. Forthcoming.

Yu, Ronald, and Gabriele Spina Alì. 2019. "What's Inside the Black Box? AI Challenges for Lawyers and Researchers." *Legal Information Management* 19: 2–13.

Zarsky, Tal Z. 2014. "Understanding Discrimination in the Scored Society." *Washington Law Review* 89: 1375–412.

Zhang, Angela Huyue. 2024. *High Wire: How China Regulates Big Tech and Governs Its Economy*. New York: Oxford University Press.

Zhang, Hongzhou, and Shaleen Khanal. 2024. "To Win the Great AI Race, China Turns to Southeast Asia." *Asia Policy* 19 (1): 21–34.

List of contributors

Mark Andrejevic is a Professor in the School of Media, Film, and Journalism at Monash University, with expertise in the social and cultural implications of data mining, and online monitoring.

Ausma Bernot is a Lecturer in Criminology in the School of Criminology and Criminal Justice at Griffith University. Her research focuses on the intersection of technology and crime, with a particular focus on surveillance and technology governance.

Rogier Creemers is an Associate Professor in the Law and Governance of China at Leiden University. His research investigates China's domestic technology policies, as well as China's participation in global cyber affairs. He is a founding member of DigiChina, a project run in cooperation with New America, as well as a frequent contributor to international news media.

Xin Dai is an Associate Professor and Vice Dean at Peking University Law School. Xin's research interests include legal theories, law and society, economic analysis of law, information privacy and internet law.

Xin Gu is an Associate Professor in the School of Media, Film, and Journalism at Monash University. She is an Expert appointed by UNESCO 2005 Convention on the Protection and Promotion of Diversity of Cultural Expression.

Haemin Jee is an Assistant Professor of International Relations at the United States Military Academy at West Point. Her research interests include China's political economy, Chinese public opinion, durability of authoritarian regimes, and democratic backsliding.

Chris O'Neill is a Research Fellow at the School of Media Film and Journalism at Monash University. He researches public responses to automated facial recognition.

Warwick Powell is Adjunct Professor at the School of Design, Queensland University of Technology. He has published numerous papers on the design and application of distributed ledger technologies and blockchains in particular, with a focus on the adoption of blockchains within China.

Neil Selwyn is a Distinguished Professor in the Faculty of Education, Monash University, with expertise in digital education.

Gavin Smith is an Associate Professor of Sociology at the Australian National University. His research has explored the social impacts of surveillance, specifically looking at the intersubjective meanings ascribed to everyday practices of watching and being watched.

Anne-Christine Trémon is full Professor at the Graduate School for Social Sciences (École des hautes études en sciences sociales, EHESS) in Paris and a member of the China–Korea–Japan Research Center. Her latest book *From Village Commons to Public Goods, Graduated Provision in Urbanizing China* was published with Berghahn in 2023.

Xuanzi Xu received her PhD at the University of Sydney in 2020. Her study focuses on how ordinary Chinese internet users' everyday news engagement contributes to the configuration of online public spheres in China.

Fan Yang is a Research Associate and PhD candidate at the University of Cologne, with a dissertation on the development of smart courts in China. She holds a bachelor's in law and a master's in human rights law from China University of Political Science and Law, and studied Law at Lund University.

Haiqing Yu is a Professor of Media and Communication and an Australian Research Council Future Fellow at RMIT University, Australia. She is a critical media studies scholar with expertise in Chinese digital media, technologies and cultures, with a focus on their sociopolitical impacts in China, Australia and the Asia Pacific.

Peter K. Yu is University Distinguished Professor and Regents Professor of Law and Communication, and Director of the Center for Law and Intellectual Property at Texas A&M University, US. He is a leading expert in international intellectual property and communications law.

Index

Printed in the United States
by Baker & Taylor Publisher Services